SOLVING
ORDINARY and PARTIAL
BOUNDARY VALUE PROBLEMS

in Science and Engineering

CRC Series in
COMPUTATIONAL MECHANICS
and APPLIED ANALYSIS

Series Editor: J.N. Reddy
Texas A&M University

New and Forthcoming Titles

APPLIED FUNCTIONAL ANALYSIS
J. Tinsley Oden and Leszek F. Demkowicz

**THE FINITE ELEMENT METHOD IN HEAT TRANSFER
AND FLUID DYNAMICS**
J.N. Reddy and D.K. Gartling

**MECHANICS OF LAMINATED COMPOSITE PLATES:
THEORY AND ANALYSIS**
J.N. Reddy

PRACTICAL ANALYSIS OF COMPOSITE LAMINATES
J.N. Reddy and Antonio Miravete

**SOLVING ORDINARY and PARTIAL BOUNDARY
VALUE PROBLEMS in SCIENCE and ENGINEERING**
Karel Rektorys

SOLVING
ORDINARY and PARTIAL
BOUNDARY VALUE PROBLEMS

in Science and Engineering

Karel Rektorys

CRC Press

Boca Raton   London   New York   Washington, D.C.

Acquiring Editor:	Bob Stern
Project Editor:	Maggie Mogck
Marketing Manager:	Jane Stark
Cover design:	Dawn Boyd

Library of Congress Cataloging-in-Publication Data

Rektorys, Karel.
 Solving ordinary and partial boundary value problems in science and engineering / Karel Rektorys.
  p. cm. -- (CRC series in computational mechanics and applied analysis)
  Includes bibliographical references and index.
  ISBN 0-8493-2552-8 (alk. paper)
  1. Science--Mathematics. 2. Engineering mathematics. 3. Differential equations--Numerical
solutions. 4. Differential equations, Partial--Numerical solutions. 5. Boundary value
problems--Numerical solutions. I. Title. II. Series.
 Q161.2.R45   1998
 620 ' .001 ' 51535—dc21

98-25766
CIP

Preface

This book has been written primarily for consumers of mathematics—for engineers and scientists (not excluding mathematicians themselves). Its aim is for readers to become familiar with ordinary and partial differential equations with boundary conditions (boundary value problems, in brief) and acquainted with fundamental properties and, especially, with efficient methods of constructing solutions or sufficiently satisfactory approximations.

The main sources for writing this book were my many years of lectures to undergraduate and postgraduate students at the Technical University in Prague, as well as to teachers of technical and physical subjects, including my younger colleagues in mathematics.

My idea, when writing the book, was to prove that even a difficult problematic can be explained in such a way that a reader, who is not a professional mathematician, can understand and solve it. This requires an appropriate explanation, of course. Moreover, I tried

1. To make a proper choice of the subject matter, preferring what is essential for the reader and neglecting useless details

2. To motivate mathematical theory by introducing adequate technical or physical examples, when useful

3. To omit some proofs that are not of a constructive character and that cannot, consequently, offer a "nonmathematical" reader a better insight into problematics (concerning, especially, proofs of some deeper results based on application of functional analysis—refer to the literature, (e.g., to my monograph [2])

4. To present special remarks for "mathematical" readers, drawing their attention to the possibility of generalization of obtained results, showing them connections between them, and so forth, to make problematics also attractive to mathematicians themselves

Each chapter ends with a section of problems, essentially of three types. Problems of the first kind are routine ones presented for practice only. Problems of the second kind are not very complicated, and can be solved by different methods, while their explicit solution is known, so that it is possible to compare numerical results and efficiency of the methods used. Problems of the third kind are more particular and have been prepared for readers who are more deeply interested in problematics. They are distinguished by an asterisk (e.g., Problem 3.7.15*).

The book is divided into five chapters. The first is devoted to ordinary differential equations with boundary conditions (including eigenvalue problems, numerical methods, etc.); the remaining chapters, to corresponding problems in partial differential equations. Individual chapters are divided into sections. If we write, for example, Section 1.5, we mean Chapter 1, Section 5. If Theorem 1.5.2 is quoted, Theorem 2 from Section 1.5 is in question. Thus, in the running head we locate number 1.5 of that section, and there we find Theorem 1.5.2.

Finally, I would like to express my sincere thanks to Professor Ivo Babuška for valuable suggestions. Further thanks are due to my colleagues Dr. V. Kelar and Dr. J. Charvát for a very careful preparation of the text—including drawings.

Karel Rektorys

Author

Karel Rektorys was born in the Czech Republic and educated in Prague (Charles' University Prague—Mathematics and Physics). Then he did theoretical research of the Škoda Works Plzeň (Pilsen). In 1961 he received Dr. Sci. in Mathematics (Academy of Sciences Prague), in 1964 became a Professor of Mathematics at the Technical University Prague, and in 1979 became a coordinator of research in applied mathematics in Czechoslovakia. In addition, he has presented many lectures at international conferences, has written monographs and articles in international journals, and has received more than 250 citations of his works in the Science Citation Index.

Most renowned publications

Mathematische Elastizitätstheorie der ebenen Probleme, with I. Babuška and F. Vyčichlo, Berlin, Akademieverlag, 1960

Variational Methods in Mathematics, Science, and Engineering, Dordrecht–Boston–London, Kluwer (Reidel), 1st ed. 1977, 2nd ed. 1979, a monograph (National Prize, 1979); translated in German (Munich–Vienna, Hanser, 1984) and in Russian (Moskva, Mir, 1985)

The Method of Discretization in Time and Partial Differential Equations, Dordrecht–Boston–London, Kluwer (Reidel), 1982, a monograph

Survey of Applicable Mathematics, 1st ed., London, Iliffe, 1969; 2nd rev. ed. (Vol. 1, 781 pp., Vol. 2, 942 pp.), Dordrecht–Boston–London, Kluwer, 1994, an encyclopedia, with co-authors, editor in chief

Main Awards and Distinctions

National Prize for Literature (1979)
State Prize for Research (1989)
Five Literary prizes
Prize of Ministry for Education (1991)
Prize of the President of Technical University Prague (1996)
Komenský Medal for Teaching (1987)
Golden Felber Medal (1988)
Golden Bolzano Medal of the Czechoslovak Academy of Sciences (1988)
Hlávka Medal for lifework (1997)
Member of editorial advisory boards
Member of the Czechoslovak Academy of Sciences (1989)
Honorary Member of the Union of Czech Mathematicians and Physicists (1987)
Honorary Member of the Česká Matice Technická (a scientific Engineering society, 1994)
Member of the American Association of Engineering Societies (AAES), 1995
His name can be found in *Who's Who* books, even in *Who's Who in Engineering,* (published by the AAES), in spite of being a mathematician, not an engineer.

Contents

Introduction

The reader is probably well acquainted with problems in ordinary differential equations with so-called *initial conditions*. An example of such a problem is the equation

$$u'' + u = 0 \tag{1}$$

with the conditions

$$u(0) = 1$$
$$u'(0) = 0 \tag{2}$$

Conditions for the solution u to be found thus are given at *one* point only.

As concerns questions on the existence of a solution, problems with initial conditions are not difficult. If the given equation is "sufficiently reasonable," then just one solution of the given problem exists. Usually it also can be easily found: Let us solve the problem (1) and (2). The general solution (= general integral) of Equation 1 is

$$u = C_1 \cos x + C_2 \sin x$$

(cf., e.g., K. Rektorys 1994 [1] Section 17.13). Conditions (2) yield the values of C_1, C_2:

$$C_1 \cos 0 + C_2 \sin 0 = 1$$

$$-C_1 \sin 0 + C_2 \cos 0 = 0$$

Thus $C_1 = 1$ and $C_2 = 0$. Consequently, the solution of the problem (1) and (2) is

$$u = \cos x$$

However, often a differential equation with so-called *boundary conditions* is to be solved. If we have to find the deflection of a bar, for example, we have to solve the corresponding differential equation with conditions characterizing how the bar is supported at its ends. (See Section 1.1.) The solution of problems with boundary conditions is more complicated. Even questions concerning the existence of a solution do not need to be simple. Let us solve, for example, the problem

$$u'' = 2 \tag{3}$$

$$u'(0) = 0$$

$$u'(1) = 0 \tag{4}$$

(Note that conditions for the solution are given at *different* points here, namely, at points $x = 0$ and $x = 1$.) Integrating Equation 3 twice, we get its general solution

$$u = x^2 + C_1 x + C_2$$

and we have to determine C_1 and C_2 to satisfy the conditions (4). However, $u' = 2x + C_1$, so that conditions (4) yield

$$2 \cdot 0 + C_1 = 0$$

$$2 \cdot 1 + C_1 = 0$$

These equations are evidently in contradiction, the first one giving $C_1 = 0$ and the second one, $C_1 = -2$. The problem (3) and (4) thus has no solution. For the reader this fact may be rather surprising, because both the equation and the boundary conditions are very simple.

On the other hand, if we solve the problem

$$u'' = 0 \tag{5}$$

$$u'(0) = 0$$

$$u'(1) = 0 \tag{6}$$

integrating Equation (5) two times, we get,

$$u = C_1 x + C_2 \tag{7}$$

Here $u' = C_1$, and both conditions (6) are satisfied with $C_1 = 0$. Then, by (7),

$$u = C_2 \tag{8}$$

where C_2 is arbitrary. Thus the problem (5) and (6) has an infinite number of solutions.

Such strange results are not encountered when solving differential equations with *initial* conditions. Thus, it may seem to the reader that problems with boundary conditions are curious problems without reasonable physical or technical motivation or interpretation. The converse is true. For example, in the case of eigenvalue problems (see Chapter 1) corresponding "unreasonabilities" are eminently interesting for engineers, producing "critical values" in buckling problems, and so forth. Generally speaking, problems with boundary conditions are, for "consumers" of mathematics, of particular interest.

Our book is devoted just to problems with boundary conditions, in ordinary as well as in partial differential equations. Let us summarize briefly the contents.

In Chapter 1, problems in ordinary differential equations with boundary conditions *(ordinary boundary value problems)* are treated, including eigenvalue problems. A suitable approximate method also is shown for the case when the differential equations considered have nonconstant coefficients (bars with variable cross sections, etc.)

In Chapter 2, the reader will become acquainted with most current problems in partial differential equations with boundary conditions *(partial boundary value problems)*. In Chapters 3 to 5, we present effective methods for solution of these (and more general) problems, in particular, variational methods, the finite-difference method, and the Fourier method.

REMARK 0.0.1

Solutions of our problems will be always denoted by u in this book, not by y, as the reader is probably accustomed to from textbooks on ordinary differential equations. The reason is simple: When a partial differential equation is considered, then its solution is a function of several variables $x, y, \ldots$, so that y has another meaning here.

Chapter 1

Ordinary Differential Equations with Boundary Conditions; Eigenvalue Problems

1.1 Technical Motivation

Let us find the deflection u of a horizontal homogeneous bar of constant cross section, length l, simply supported at its ends, and loaded vertically by a uniformly distributed load q per unit length (the self-weight of the bar being neglected) and axially by a compressive force P (see Figure 1.1.1, where the reactions R_1 and R_2 as well as the orientation of coordinate axes x, u, also are shown). Let E be the Young modulus of elasticity and I, the moment of inertia (= the second moment of area) of the cross section with respect to the axis of the bar. By taking the orientation of the coordinate axes into account, the deflection $u(x)$ at an arbitrary point $A(x, u)$ satisfies the equation (as well known from the theory of elasticity)

$$u'' = -\frac{M}{EI} \tag{1.1.1}$$

where M is the "left-hand" bending moment with respect to the (already displaced) point $A(x, u)$. Thus M is the sum of three moments, induced by the reaction R_1, by the force P, and by the part of the vertical loading in the interval $[0, x]$

$$M = +\frac{ql}{2} \cdot x + P \cdot u - qx \cdot \frac{x}{2}$$

Differential Equation 1.1.1 then becomes

$$u'' = -\frac{1}{EI}\left(Pu + \frac{qlx}{2} - \frac{qx^2}{2}\right)$$

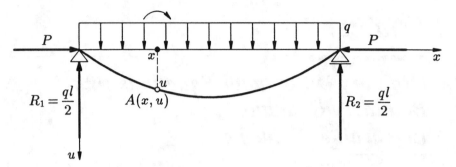

FIGURE 1.1.1

If we denote

$$\frac{P}{EI} = a^2 \quad (a > 0)$$

$$\frac{q}{2EI} = b \tag{1.1.2}$$

$$-\frac{ql}{2EI} = c$$

we have

$$u'' + a^2 u = bx^2 + cx \tag{1.1.3}$$

This is the differential equation for the deflection of the bar to be found. To this equation, boundary conditions

$$u(0) = 0$$

$$\tag{1.1.4}$$

$$u(l) = 0$$

are to be added, because the deflection is zero at the endpoints of the bar. (Let us note that, in general, an equation of the fourth order is to be considered with the view of being able to characterize, by a sufficient number of boundary conditions, the kind of support of the bar at its endpoints. However, here Equations (1.1.3) and (1.1.4) yield

$$u''(0) = -a^2 u(0) + b \cdot 0^2 + c \cdot 0 = 0$$

$$u''(l) = -a^2 u(l) + bl^2 + cl = 0$$

because $bl^2 + cl = 0$ according to the last relations in Equation (1.1.2). It means that the moments at the endpoints of the bar are zero, and this is just what characterizes a simply supported bar. So here it is sufficient to consider second-order Equation (1.1.3) with the conditions (1.1.4) only.)

Let us solve the problem (1.1.3) and (1.1.4). The general solution of Equation (1.1.3) is of the form

$$u = u_p + u_h \tag{1.1.5}$$

where u_p is a function satisfying Equation (1.1.3) (the so-called particular solution of Equation (1.1.3) and u_h is the general solution of the corresponding homogeneous equation

$$u'' + a^2 u = 0 \tag{1.1.6}$$

see, e.g., K. Rektorys 1994, [1] Section 17.11. However, the general solution of Equation (1.1.6) is well known (see K. Rektorys 1994, [1] Section 17.13)

$$u_h = C_1 \cos ax + C_2 \sin ax \tag{1.1.7}$$

The function u_p can be found in the form

$$u_p = Ax^2 + Bx + C \tag{1.1.8}$$

(K. Rektorys 1994, [1] Section 17.14), where A, B, C are to be determined so that u_p satisfies Equation (1.1.3). Putting $Ax^2 + Bx + C$ for u and $2A$ for u'' into Equation (1.1.3), we get

$$2A + a^2(Ax^2 + Bx + C) = bx^2 + cx$$

Compare coefficients at the same powers of x to obtain

$$A = \frac{b}{a^2}$$

$$B = \frac{c}{a^2}$$

$$C = -\frac{2A}{a^2} = -\frac{2b}{a^4}$$

So the general solution of Equation (1.1.3) is (by (1.1.5), (1.1.7), and (1.1.8))

$$u = \frac{b}{a^2} x^2 + \frac{c}{a^2} x - \frac{2b}{a^4} + C_1 \cos ax + C_2 \sin ax$$

The first of boundary conditions (1.1.4) gives (thus putting $x = 0$, $u = 0$)

$$0 = \frac{b}{a^2} \cdot 0^2 + \frac{c}{a^2} \cdot 0 - \frac{2b}{a^4} + C_1 \cdot 1 + C_2 \cdot 0$$

from which

$$C_1 = \frac{2b}{a^4}$$

Thus

$$u = \frac{b}{a^2} x^2 + \frac{c}{a^2} x + \frac{2b}{a^4} (\cos ax - 1) + C_2 \sin ax$$

The second condition (1.1.4) yields (putting $x = l$, $u = 0$)

$$0 = \frac{b}{a^2} l^2 + \frac{c}{a^2} l + \frac{2b}{a^4} (\cos al - 1) + C_2 \sin al \qquad (1.1.9)$$

$$= \frac{2b}{a^4} (\cos al - 1) + C_2 \sin al$$

since

$$\frac{bl^2}{a^2} + \frac{cl}{a^2} = 0$$

(because of the last relations in (1.1.2), see earlier).

Thus if

$$\sin al \neq 0 \qquad (1.1.10)$$

we obtain

$$C_2 = \frac{2b}{a^4} \frac{1 - \cos al}{\sin al}$$

and

$$u = \frac{b}{a^2} x^2 + \frac{c}{a^2} x + \frac{2b}{a^4} (\cos ax - 1) + \frac{2b}{a^4} \frac{1 - \cos al}{\sin al} \sin ax \qquad (1.1.11)$$

where l is the length of the bar and a, b, and c are given by (1.1.2). This is the deflection of the bar considered.

The obtained solution can be put to the form from which it can be seen that the deflection is symmetrical with respect to line $x = l/2$, which could be expected directly from the beginning. We do not go into detail. What is much more interesting is the following fact: If

$$al = \pi \qquad (1.1.12)$$

the given problem is not solvable. In fact, then $\sin al = 0$, $\cos al = -1$, and Equation (1.1.9) becomes

$$0 = -\frac{4b}{a^4} + C_2 \cdot 0$$

which cannot be fulfilled (because of $b \neq 0$) for any C_2. Thus the second of boundary conditions (1.1.4) cannot be satisfied; consequently, the problem (1.1.3) and (1.1.4) has no solution.

This fact is interesting, first, from the mathematical point of view. From the point of view of the theory of elasticity, we have the following phenomenon: From (1.1.12) it follows

$$a^2 = \frac{\pi^2}{l^2}$$

and according to the first relation in (1.1.2) we have

$$P = \frac{EI\pi^2}{l^2}$$

This value of P is usually called *critical value* of this force *(critical force)* or *buckling load,* and is denoted by P_b

$$P_b = \frac{EI\pi^2}{l^2} \tag{1.1.13}$$

For a technical interpretation, first, let P be smaller than the buckling load, that is, let $P < P_b$. Then $0 < al < \pi$ and we get a unique deflection by (1.1.11). Now, if $P \to P_b$, then $al \to \pi$, $\sin al \to 0$, remaining positive. All terms in (1.1.11) remain bounded, except the last one. If we take, for example, the middle point $x = l/2$ of the bar, we get

$$\sin ax = \sin \tfrac{al}{2} \doteq \sin \tfrac{\pi}{2} = 1$$

$$1 - \cos al \doteq 1 - \cos \pi = 2$$

so that, because of $\sin al \to 0+$, the last term in (1.1.11) tends to infinity. The deflection becomes "infinite," and the bar fails. See also Problem 1.10.7 for another vertical loading of the bar.

1.2 Relation Between the Problem (1.1.3) and (1.1.4) and the Problem Corresponding to the Zero Vertical Loading of the Bar

Let us note (see 1.1.13) that the buckling load P_b does not depend on the vertical loading of the bar (to be able to determine C_2 from (1.1.9), it is essential to know whether $\sin al \neq 0$ or $\sin al = 0$). Consider now the case with zero vertical load, that is, from the mathematical point of view, the problem is

$$u'' + a^2 u = 0 \tag{1.2.1}$$

$$u(0) = 0$$
$$\tag{1.2.2}$$
$$u(l) = 0$$

Intuitively there is no reason why the bar should deflect. However, here the buckling load plays its role: The general solution of (1.2.1) is

$$u = C_1 \cos ax + C_2 \sin ax$$

The first condition in (1.2.2) yields $C_1 = 0$; thus

$$u = C_2 \sin ax$$

The second one gives

$$0 = C_2 \sin al \tag{1.2.3}$$

If $0 < P < P_b$, then $0 < al < \pi$. Thus $\sin al > 0$ and $C_2 = 0$ by (1.2.3). We get the unique solution

$$u \equiv 0$$

If $P = P_b$, we have $al = \pi$, $\sin al = 0$, and C_2 in (1.2.3) is arbitrary. In this case, the problem (1.2.1) and (1.2.2) has an infinite number of solutions

$$u = C_2 \sin ax \tag{1.2.4}$$

$$C_2 \text{ arbitrary}$$

Thus if $P = P_b$, we have, in the case of nonzero vertical loading an "infinite" deflection and in the case of the zero loading an "indefinite" deflection, and the bar is "unstable." The second case is important especially when the bar plays the role of a column. The bar is then in vertical position, the compressive force is vertical as well (the bar carries a construction), and horizontal loading is equal to zero. If the weight of the construction, acting on the column, attains the value of P_b, the horizontal deflection may remain equal to zero (according to (1.2.4), for $C_2 = 0$); but a small horizontal load (e.g., small gust of wind) is sufficient to convert the situation into that considered in Section 1.1, leading to an infinite (here horizontal) deflection, and the column fails.

Let us note that difficulties with solvability of the problem (1.1.3) and (1.1.4) arise exactly if a^2 is such that the homogeneous problem (1.2.1) and (1.2.2) also has solutions other than the zero one. The relation between problems of the type found in (1.1.3) and (1.1.4) (thus problems with a nonhomogeneous differential equation) and (1.2.1) and (1.2.2) (with the corresponding homogeneous equation) is interesting from the technical as well as the mathematical point of view (even in much more complicated problems than ours) and leads in a natural way to so-called eigenvalue problems. We shall discuss these problems in Section 1.4. First, we recall basic concepts concerning the space $L_2(a,b)$, which will be a powerful tool for us throughout the following sections.

1.3 Brief Summary of the Space $L_2(a,b)$

The reader probably has some knowledge of the so-called *space $L_2(a,b)$* and of orthogonal systems and Fourier series in this space. However, for convenience, here we summarize basic concepts concerning these problematics. (For more details see, e.g., K. Rektorys 1994, [1] Chapter 16).

Thus let us recall that the space $L_2(a,b)$ is a linear space, the elements of which are square integrable functions[1] in the (bounded) interval $[a, b]$ (thus the integrals

$$\int_a^b u(x)\,dx$$

$$\int_a^b u^2(x)\,dx$$

---

[1]Square integrable in the so-called *Lebesgue sense:* The assumption on integrability in this sense is essential from the point of view of mathematical theory, but for an engineer or scientist it is not as important from the point of view of applications. Thus we do not go into detail here. In a surveyable form, it is possible to obtain a certain knowledge of the Lebesgue integral in K. Rektorys 1994, [1] Section 13.14; or in K. Rektorys 1980, [2] Chapter 28.

exist and are finite), while in this space the *scalar product* (u, v), the *norm* $\|u\|$, and the *distance* (or *metrics*) $\rho(u, v)$ are defined by the relations

$$(u, v) = \int_a^b uv\,dx \qquad (1.3.1)$$

$$\|u\| = \sqrt{(u, u)} \qquad (1.3.2)$$

$$\rho(u, v) = \|u - v\| \qquad (1.3.3)$$

Two functions for which

$$\rho(u, v) = 0 \qquad (1.3.4)$$

that is

$$\|u - v\| = 0$$

holds, are called *equivalent in the space* $L_2(a,b)$; we write

$$u = v \quad \text{in } L_2(a,b) \qquad (1.3.5)$$

All mutually equivalent functions are considered as one element of the space $L_2(a,b)$. Equivalent functions may differ, in the interval $[a, b]$, only on a so-called set of measure zero, for example, at a finite number of points. (We also say that they are equal in the interval $[a, b]$ *almost everywhere*.) If two functions are equivalent and continuous in $[a, b]$, then they are equal in the whole interval $[a, b]$, that is, then it holds

$$u = v \text{ in } L_2(a,b) \iff u(x) = v(x) \text{ for all } x \in [a, b] \qquad (1.3.6)$$

Especially, if u is a continuous function in $[a, b]$, then

$$u = 0 \text{ in } L_2(a,b) \iff u(x) = 0 \text{ for all } x \in [a, b] \qquad (1.3.7)$$

Example 1.3.1
For the scalar product of the functions

$$u = \sin x$$

$$v = \cos x$$

in $L_2(0, \pi/2)$ (i.e., on the interval $[0, \pi/2]$) we have by Equation 1.3.1

$$(u, v) = \int_0^{\pi/2} \sin x \cos x \, dx = \left[\frac{\sin^2 x}{2}\right]_0^{\pi/2} = \frac{1}{2}$$

in the space $L_2(0, \pi)$ (i.e., on the interval $[0, \pi]$)

$$(u, v) = \int_0^{\pi} \sin x \cos x \, dx = \left[\frac{\sin^2 x}{2}\right]_0^{\pi} = 0$$

It is evident that the value of the scalar product depends on the interval on which this scalar product is calculated (i.e., on the space in which we calculate it). If there is a danger of misunderstanding, we write in more detail

$$(u, v)_{L_2(a,b)}$$

that is,

$$(u, v)_{L_2(0,\pi/2)}$$

$$(u, v)_{L_2(0,\pi)}$$

and so forth.   ⬚

Example 1.3.2

The norm of the function $u = \sin x$ in $L_2(0, \pi)$ is equal to $\sqrt{\pi/2}$, because by Equation 1.3.2 we have

$$\|u\|^2 = (u, u) = \int_0^{\pi} u^2 dx$$

$$= \int_0^{\pi} \sin^2 x dx = \int_0^{\pi} \frac{1 - \cos 2x}{2} dx = \frac{1}{2}\left[x - \frac{\sin 2x}{2}\right]_0^{\pi} = \frac{\pi}{2}$$

thus

$$\|u\| = \sqrt{\pi/2}$$

⬚

Example 1.3.3
The functions

$$u(x) \equiv 0$$

$$v(x) = \begin{cases} 0 \text{ for } x \in [0, 1) \\ 1 \text{ for } x = 1 \end{cases}$$

are equivalent functions in $L_2(0, 1)$ (they differ only at one point, $x = 1$).   ☐

Let us recall basic properties of the scalar product:

$$(u, v) = (v, u) \qquad \text{(symmetry)} \tag{1.3.8}$$

$$(a_1 u_1 + a_2 u_2, v) = a_1 (u_1, v) + a_2 (u_2, v) \quad \text{(additivity)} \tag{1.3.9}$$

$$(u, u) \overset{\geq}{=} 0 \tag{1.3.10}$$

$$(u, u) = 0 \iff u = 0 \text{ in } L_2(a,b) \tag{1.3.11}$$

We say that we *multiply* the equation

$$u = f \tag{1.3.12}$$

scalarly by the function v if we multiply it by v and then integrate over the interval $[a, b]$

$$\int_a^b uv \, dx = \int_a^b fv \, dx$$

The result is then written in the form

$$(u, v) = (f, v) \tag{1.3.13}$$

(therefore, "scalar multiplication").
   A function u is called *normed* in the space $L_2(a,b)$ if its norm is equal to unity. Two functions u and v are called *orthogonal* in $L_2(a,b)$, and we write

$$u \perp v \quad \text{in } L_2(a,b) \tag{1.3.14}$$

if

$$(u, v) = 0$$

Example 1.3.4
The function

$$u = \sqrt{\frac{2}{\pi}} \sin x$$

is a normed function in $L_2(0, \pi)$, because

$$\|u\|^2 = \frac{2}{\pi} \int_0^\pi \sin^2 x \, dx = 1$$

(see Example 1.3.2).
  The functions

$$u = \sin x,$$

$$v = \cos x$$

are orthogonal in the space $L_2(0, \pi)$, since

$$(u, v) = \int_0^\pi \sin x \cos x \, dx = 0$$

(see Example 1.3.1). They are not orthogonal, for example, in the space $L_2(0, \pi/2)$, since in that space we have

$$(u, v) = \int_0^{\pi/2} \sin x \cos x \, dx = \frac{1}{2}$$

(see the same example). □

  A system of functions

$$u_1, u_2, \ldots \tag{1.3.15}$$

(here it is more usual to speak about a system than about a sequence) is called orthogonal in the space $L_2(a,b)$, if every two functions u_i and u_j of this system, $i \neq j$, are orthogonal in that space. If, moreover, every function of this system is normed, the system in (1.3.15) is called *orthonormalized* or *orthonormal* in that space.

Example 1.3.5
The system of functions

$$\sin x, \quad \sin 2x, \quad \ldots, \quad \sin nx, \quad \ldots \tag{1.3.16}$$

is orthogonal in the space $L_2(0, \pi)$, because for every two functions $\sin ix$ and $\sin jx$ where $i \neq j$, we have

$$(\sin ix, \sin jx) = \int_0^\pi \sin ix \sin jx \, dx = 0$$

(cf. Remark 1.5.1 and Problem 1.10.9).

If we multiply each function of the system (1.3.16) by the number $\sqrt{2/\pi}$, we obtain an orthonormal system. ☐

A sequence of functions

$$v_1, v_2, \ldots \tag{1.3.17}$$

is said to *converge* (= to be *convergent*) in the space $L_2(a,b)$, or *in the mean on the interval* $[a, b]$ (*in the mean*, in brief) and to have the *limit* v, if

$$\lim_{n \to \infty} \rho(v, v_n) = 0 \tag{1.3.18}$$

We write

$$\lim_{n \to \infty} v_n = v \quad \text{in } L_2(a,b) \tag{1.3.19}$$

According to (1.3.1) to (1.3.3), the limit (1.3.18) means that

$$\lim_{n \to \infty} \sqrt{\int_a^b (v - v_n)^2 \, dx} = 0$$

or, what is the same, that

$$\lim_{n \to \infty} \int_a^b (v - v_n)^2 \, dx = 0 \tag{1.3.20}$$

We say that the *series*

$$v_1 + v_2 + \cdots \tag{1.3.21}$$

converges (= is convergent) in the space $L_2(a,b)$ (= in the mean on the interval $[a, b]$) and *has the sum s*, if the sequence of its partial sums

$$s_n = \sum_{k=1}^n v_k$$

is convergent in the space $L_2(a,b)$ (= in the mean) and has the limit s. We write

$$\sum_{n=1}^{\infty} v_n = s \quad \text{in } L_2(a,b) \tag{1.3.22}$$

Example 1.3.6

The sequence of functions

$$v_n(x) = x^n$$

$$n = 1, 2, \ldots$$

converges in the space $L_2(0, 1)$ (in the mean on the interval $[0, 1]$) to the zero function $v = 0$, because (see (1.3.20))

$$\lim_{n\to\infty} \int_0^1 (0 - x^n)^2 \, dx = \lim_{n\to\infty} \int_0^1 x^{2n} \, dx = \lim_{n\to\infty} \frac{1}{2n+1} = 0$$

Let us note that, at the same time, the given sequence converges pointwise in the interval $[0, 1]$ to the function

$$v(x) = \begin{cases} 0 & \text{for } x \in [0, 1) \\ 1 & \text{for } x = 1 \end{cases}$$

since for every $x \in [0, 1)$ we have

$$\lim_{n\to\infty} x^n = 0$$

and for $x = 1$

$$\lim_{n\to\infty} 1^n = 1$$

However, from the point of view of the space $L_2(0, 1)$ this result is not different from that obtained earlier, because (see Example 1.3.3) the functions $v(x)$ and the zero function are equivalent in that space.   ☐

An orthonormal system (1.3.15) is called *complete* (or an *orthonormal basis*) in the space $L_2(a,b)$, if every function $u \in L_2(a,b)$ can be expressed as the sum, in the mean, of the series

$$b_1 u_1 + b_2 u_2 + \cdots \tag{1.3.23}$$

where

$$b_n = (u, u_n)$$

$$n = 1, 2, \ldots$$

The series (1.3.23) is often called the *Fourier series*, or *Fourier expansion* of the function u with respect to the orthonormal system (1.3.15). To prove completeness of a system is not easy, in general.

Example 1.3.7
A typical orthonormal basis in $L_2(0, \pi)$ is the system of functions

$$v_1 = \sqrt{\frac{2}{\pi}} \sin x$$

$$v_2 = \sqrt{\frac{2}{\pi}} \sin 2x, \ldots$$

☐

Example 1.3.8
For the function

$$u(x) \equiv 1 \quad \text{in } [0, \pi]$$

we have

$$b_n = \int_0^\pi 1 \cdot \sqrt{\frac{2}{\pi}} \sin nx \, dx = -\sqrt{\frac{2}{\pi}} \left[\frac{\cos nx}{n} \right]_0^\pi = \begin{cases} \sqrt{\frac{2}{\pi}} \cdot \frac{2}{n} & \text{for } n \text{ odd} \\ 0 & \text{for } n \text{ even} \end{cases}$$

thus

$$1 = \frac{4}{\pi} \left(\sin x + \frac{\sin 3x}{3} + \frac{\sin 5x}{5} + \cdots \right) \quad \text{in } L_2(0, \pi)$$

☐

1.4 Eigenvalue Problems

For technical motivation see Sections 1.1 and 1.2.

Consider the problem

$$u'' + \lambda u = 0 \qquad (1.4.1)$$

$$u(0) = 0 \qquad (1.4.2)$$

$$u(l) = 0$$

where λ is a real parameter. Under a solution of this problem, for a fixed $\lambda \in (-\infty, +\infty)$, we understand such a function u that is continuous in the interval $[0, l]$ satisfies the conditions in (1.4.2), and in the open interval $(0, l)$ has two continuous derivatives and fulfills Equation 1.4.1 there.

For every λ, the problem (1.4.1) and (1.4.2) evidently has the solution $u \equiv 0$. This function, thus the function

$$u(x) = 0 \quad \text{for every } x \in [0, l]$$

is called the *zero solution*, or *trivial solution* of that problem. However, for some values of λ there also exists a nonzero solution of that problem (thus a solution that is not zero identically). For example, if

$$\lambda = \frac{\pi^2}{l^2} \qquad (1.4.3)$$

then the function

$$u = \sin \frac{\pi x}{l} \qquad (1.4.4)$$

is a solution of the given problem. In fact, the function (1.4.4) is continuous in the interval $[0, l]$, is equal to zero for $x = 0$ and $x = l$ and has two continuous derivatives in the interval $(0, l)$, (even in the interval $(-\infty, +\infty)$) while

$$u'' = -\frac{\pi^2}{l^2} \sin \frac{\pi x}{l}$$

so that

$$u'' + \frac{\pi^2}{l^2} u = 0$$

DEFINITION 1.4.1   (*eigenvalues and eigenfunctions of the problem (1.4.1) and (1.4.2)*).

Such value of the parameter λ for which the problem (1.4.1) and (1.4.2) has, except the zero solution, also a nonzero one, is called an *eigenvalue* of that problem. The nonzero solution u in question is called the *eigenfunction* corresponding to that λ.

Thus (1.4.3) is an eigenvalue and the function (1.4.4) the corresponding eigenfunction of the problem (1.4.1) and (1.4.2).

It follows directly from the definition that an eigenfunction u is a continuous function in the interval $[0, l]$, not identically equal to zero, and, consequently, its norm in $L_2(0, l)$ is positive,

$$\|u\| > 0 \qquad (1.4.5)$$

Further, if u is an eigenfunction corresponding to an eigenvalue λ, then the function Cu, where $C \neq 0$, is again an eigenfunction corresponding to the same λ. Because Cu is again a nonzero continuous function in $[0, l]$, equal to zero at the endpoints of the interval $[0, l]$, and satisfies the given equation again, since that equation is linear and homogeneous. For example, the function

$$u = 3 \sin \frac{\pi x}{l}$$

is another eigenfunction corresponding to the eigenvalue in (1.4.3).

DEFINITION 1.4.2   The problem of finding all eigenvalues and eigenfunctions of (1.4.1) and (1.4.2) is called the *eigenvalue problem*.

REMARK 1.4.1

In the case of such a simple problem as that of (1.4.1) and (1.4.2), the eigenvalues and corresponding eigenfunctions can be found easily by direct computation (see Section 1.5). In spite of this, we first formulate two theorems that will be immediately useful for us, and whose almost literal analogy we meet in much more complicated cases (equations of higher orders, partial differential equations). Because of their fundamental importance, we also give corresponding proofs that are very intuitive here and give, at the same time, a good insight into the whole of the problematics.

1.5   Basic Properties of Eigenvalues and Eigenfunctions

THEOREM 1.5.1 *(on positivity of eigenvalues)*.
The problem (1.4.1) and (1.4.2) may have only positive eigenvalues.

PROOF   Let λ be an eigenvalue and u the corresponding eigenfunction, so that we have

$$u'' + \lambda u = 0 \tag{1.5.1}$$

$$u(0) = 0 \tag{1.5.2}$$

$$u(l) = 0$$

with

$$\|u\| > 0 \tag{1.5.3}$$

(see (1.4.5)). Multiply (1.5.1) scalarly (cf. (1.3.13)) by the function u. We obtain

$$(u'', u) + \lambda(u, u) = 0 \tag{1.5.4}$$

Now, integrating by parts and using the conditions in (1.5.2), we get

$$(u'', u) = \int_0^l u'' u \, dx = \left[u'u\right]_0^l - \int_0^l u'u' \, dx = -\int_0^l u'^2 \, dx \leqq 0 \tag{1.5.5}$$

Further

$$(u, u) = \|u\|^2 > 0 \tag{1.5.6}$$

by (1.5.3). Thus, by using (1.5.4)

$$\lambda = -\frac{(u'', u)}{(u, u)} \stackrel{\geq}{\leq} 0 \tag{1.5.7}$$

Equation (1.5.7) shows that if λ is an eigenvalue of the problem (1.5.1) and (1.5.2), then it must be positive, or zero. To complete the proof, we have to show that it cannot be equal to zero. Thus let $\lambda = 0$. Then the problem (1.5.1) and (1.5.2) becomes

$$u'' = 0 \tag{1.5.8}$$

$$u(0) = 0 \tag{1.5.9}$$

$$u(l) = 0$$

Integrating two times, we get the general solution of (1.5.8)

$$u = C_1 x + C_2$$

The boundary conditions (1.5.2) then yield, subsequently, $C_2 = 0$, $C_1 = 0$, so that

$$u \equiv 0$$

Consequently, $\lambda = 0$ cannot be an eigenvalue, because then the problem (1.5.1) and (1.5.2) has no nonzero solution. This completes the proof of Theorem 1.5.1.

THEOREM 1.5.2 *(on orthogonality of eigenfunctions).*
Let λ_i and λ_j be two eigenvalues of the problem (1.5.1) and (1.5.2), let u_i and u_j be corresponding eigenfunctions. If $\lambda_i \neq \lambda_j$, then the functions u_i and u_j are orthogonal in $L_2(0, l)$.

In brief
$$\lambda_i \neq \lambda_j \implies (u_i, u_j) = 0$$

PROOF   Let λ_i, or λ_j be eigenvalues of the problem (1.5.1) and (1.5.2) and u_i, or u_j corresponding eigenfunctions, respectively, so that we have

$$u_i'' + \lambda_i u_i = 0 \tag{1.5.10}$$

$$u_j'' + \lambda_j u_j = 0 \tag{1.5.11}$$

while u_i and u_j are nonzero functions fulfilling

$$u_i(0) = 0 \tag{1.5.12}$$

$$u_i(l) = 0$$

$$u_j(0) = 0 \tag{1.5.13}$$

$$u_j(l) = 0$$

Multiply (1.5.10) scalarly by the function u_j and (1.5.11) scalarly by the function u_i to get

$$(u_i'', u_j) + \lambda_i(u_i, u_j) = 0 \tag{1.5.14}$$

$$(u_j'', u_i) + \lambda_j(u_j, u_i) = 0 \tag{1.5.15}$$

However, $u_j(0) = 0$, $u_j(l) = 0$, thus

$$(u_i'', u_j) = \int_0^l u_i'' u_j \, dx = \left[u_i' u_j \right]_0^l - \int_0^l u_i' u_j' \, dx = - \int_0^l u_i' u_j' \, dx = -(u_i', u_j')$$

and similarly

$$(u_j'', u_i) = -(u_j', u_i')$$

Symmetry of the scalar product implies

$$(u_j', u_i') = (u_i', u_j')$$

$$(u_j, u_i) = (u_i, u_j)$$

Thus Equations 1.5.14 and 1.5.15 can be written in the form

$$-(u_i', u_j') + \lambda_i(u_i, u_j) = 0 \qquad (1.5.16)$$

$$-(u_i', u_j') + \lambda_j(u_i, u_j) = 0 \qquad (1.5.17)$$

Subtracting Equation 1.5.16 from Equation 1.5.17, we obtain

$$(\lambda_j - \lambda_i)(u_i, u_j) = 0$$

which yields the wanted result

$$(u_i, u_j) = 0$$

because of the assumption $\lambda_i \neq \lambda_j$.

With the help of Theorem 1.5.1, we now calculate eigenvalues and eigenfunctions of the problem (1.5.1) and (1.5.2).

By this theorem, the problem (1.5.1) and (1.5.2) can have only positive eigenvalues. This result essentially facilitates their computation. In fact, we may put

$$\lambda = k^2 \qquad (1.5.18)$$

with

$$k > 0 \qquad (1.5.19)$$

The problem (1.5.1) and (1.5.2) then becomes

$$u'' + k^2 u = 0 \tag{1.5.20}$$

$$u(0) = 0 \tag{1.5.21}$$

$$u(l) = 0$$

The general solution of (1.5.20) is (K. Rektorys 1994, [1] Section 17.13)

$$u = A_1 \cos kx + A_2 \sin kx$$

The first condition (1.5.21) gives

$$0 = A_1 \cdot 1 + A_2 \cdot 0$$

from which $A_1 = 0$ and, consequently

$$u = A_2 \sin kx \tag{1.5.22}$$

The second condition (1.5.21) then yields

$$0 = A_2 \sin kl \tag{1.5.23}$$

Now $\lambda = k^2$ is to be an eigenvalue according to the assumption. It implies $A_2 \neq 0$. Otherwise the function (1.5.22) would be identically equal to zero and, consequently, could not be an eigenfunction. Equation (1.5.23) then yields

$$\sin kl = 0 \tag{1.5.24}$$

This condition is fulfilled exactly if kl is equal to $n\pi$, where n is an integer. With regard to $l > 0$ and $k > 0$ by (1.5.19), we have to consider only positive n, $n = 1, 2, \ldots$, and the condition

$$kl = n\pi$$

then yields a countable set of values for k

$$k_1 = \frac{\pi}{l}, \; k_2 = \frac{2\pi}{l}, \; k_3 = \frac{3\pi}{l}, \; \ldots$$

Thus by (1.5.18) we obtain a countable set of eigenvalues

$$\lambda_1 = \frac{\pi^2}{l^2}, \ \lambda_2 = \frac{4\pi^2}{l^2}, \ \lambda_3 = \frac{9\pi^2}{l^2}, \ \ldots \tag{1.5.25}$$

of the problem (1.5.1) and (1.5.2). To every λ_n $(n = 1, 2, 3, \ldots)$ there correspond, according to (1.5.22) (where $A_2 \neq 0$ is arbitrary), an infinite number of eigenfunctions

$$u_n = C_n \sin \frac{n\pi x}{l} \tag{1.5.26}$$

$$C_n \neq 0 \text{ arbitrary}$$

REMARK 1.5.1

It is convenient to work with normed eigenfunctions (whose norms are equal to one). To each of the eigenvalues (1.5.25) we assign exactly one normed eigenfunction

$$v_n = \sqrt{\frac{2}{l}} \sin \frac{n\pi x}{l} \tag{1.5.27}$$

According to Theorem 1.5.2, every two functions v_i and v_j, corresponding to different eigenvalues λ_i and λ_j, are orthogonal in $L_2(0, l)$. (For direct verification of this property see Problem 1.10.9.) Thus the system of functions (1.5.27) is an orthonormal system in $L_2(0, l)$. Moreover, this system can be proved to be complete, and, consequently, it is an orthonormal basis in $L_2(0, l)$ (cf. Section 1.3). Let us summarize the obtained results into the theorem that follows.

THEOREM 1.5.3 *(summarizing theorem on the eigenvalue problem (1.5.1) and (1.5.2)).*
Consider the problem

$$u'' + \lambda u = 0 \tag{1.5.28}$$

$$u(0) = 0 \tag{1.5.29}$$

$$u(l) = 0$$

Then:

1. There exists a countable set of eigenvalues, each of them being positive

$$\lambda_n = \frac{n^2 \pi^2}{l^2} \tag{1.5.30}$$

$$n = 1, 2, \ldots$$

Obviously $\lambda_n \to +\infty$ for $n \to \infty$.

2. *To every eigenvalue λ_n there correspond an infinite number of eigenfunctions*

$$u_n = C_n \sin \frac{n\pi x}{l} \tag{1.5.31}$$

$C_n \neq 0$ *arbitrary*

3. *Eigenfunctions corresponding to different eigenvalues are orthogonal in $L_2(0, l)$.*

4. *The system of normed eigenfunctions*

$$v_n = \sqrt{\frac{2}{l}} \sin \frac{n\pi x}{l} \tag{1.5.32}$$

$$n = 1, 2, \ldots$$

constitutes an orthonormal basis in $L_2(0, l)$.

REMARK 1.5.2   *(eigenvalue problems on the interval $[a, b]$).*

In the preceding text, we considered the interval $[0, l]$, just because this interval is mostly encountered in applications. For an interval $[a, b]$ (thus for (1.5.28) with boundary conditions $u(a) = 0$ and $u(b) = 0$), Theorem 1.5.3 remains valid, only instead of the space $L_2(0, l)$ we consider the space $L_2(a,b)$ and instead of eigenvalues (1.5.30) and eigenfunctions (1.5.31) or (1.5.32), we have to consider the eigenvalues

$$\lambda_n = \frac{n^2 \pi^2}{(b - a)^2} \tag{1.5.33}$$

$$n = 1, 2, \ldots$$

and the eigenfunctions

$$u_n = C_n \sin \frac{n\pi(x - a)}{b - a} \tag{1.5.34}$$

$C_n \neq 0$ *arbitrary*

or

$$v_n = \sqrt{\frac{2}{b-a}} \, \sin \frac{n\pi(x-a)}{b-a} \qquad (1.5.35)$$

$$n = 1, 2, \ldots$$

respectively.

REMARK 1.5.3

For eigenvalue problems concerning equations of the fourth order see (K. Rektorys 1994, [1], Section 17.21, Equation 162). For generalization to equations of an arbitrary even order see, for example, K. Rektorys 1994, [1] Section 17.17. See also K. Rektorys 1980, [2] Part V; and the present book, Section 3.6, and also Problems 1.10.11 and 1.10.12.

1.6 Nonhomogeneous Equations with Boundary Conditions

In Section 1.1, we encountered a problem of the form

$$u'' + \lambda u = f \qquad (1.6.1)$$

$$u(0) = 0 \qquad (1.6.2)$$

$$u(l) = 0$$

where we had $\lambda = a^2$ and $f(x) = bx^2 + cx$. We proved that, if the parameter λ attained a certain "critical" value, then the given problem was not solvable, while the problem

$$u'' + \lambda u = 0 \qquad (1.6.3)$$

$$u(0) = 0 \qquad (1.6.4)$$

$$u(l) = 0$$

had just for that critical value, except the zero solution, also nonzero solutions.

As we shall see, this connection between the problems (1.6.1), (1.6.2), (1.6.3), and (1.6.4) is not accidental. So let us consider both of these problems with λ real and arbitrary (thus not necessarily positive), and with f continuous in $[0, l]$. The problem (1.6.3) and (1.6.4) is called the *homogeneous problem corresponding to the problem* (1.6.1) *and* (1.6.2). Thus we have the theorem that follows.

THEOREM 1.6.1 *(on solvability of the problem (1.6.1) and (1.6.2)).*

1. Let λ in (1.6.1) not be an eigenvalue of the problem (1.6.3) and (1.6.4), thus let

$$\lambda \neq \frac{n^2\pi^2}{l^2} \tag{1.6.5}$$

$$n = 1, 2, \ldots$$

Then for every function f, continuous in the interval $[0, l]$, the problem (1.6.1) and (1.6.2) has exactly one solution.

2. Let λ in (1.6.1) be equal to an eigenvalue of the problem (1.6.3) and (1.6.4); thus let

$$\lambda = \frac{n^2\pi^2}{l^2} \quad \textit{for a positive integer } n \tag{1.6.6}$$

Then the problem (1.6.1) and (1.6.2) is not solvable, in general. A necessary condition for solvability of (1.6.1) and (1.6.2) is that f be orthogonal, in $L_2(0, l)$, to every eigenfunction

$$u_n = C_n \sin \frac{n\pi x}{l}$$

$$C_n \neq 0 \textit{ arbitrary}$$

corresponding to that eigenvalue, that is, that

$$(f, u_n) = 0 \quad \textit{for every } C_n \neq 0 \tag{1.6.7}$$

or, which is the same, that

$$\left(f, \sin \frac{n\pi x}{l}\right) = 0 \tag{1.6.8}$$

be fulfilled.

If the condition (1.6.8) is satisfied, then the problem (1.6.1) and (1.6.2) is actually solvable, and there is even an infinite number of solutions.

We do not give the PROOF of Theorem 1.6.1 here (it is rather difficult and is given, for an essentially more general case, e.g., in K. Rektorys 1980, [2] Chapter 39). Here we show only what seems to be most interesting for the reader, namely, that (1.6.7) or (1.6.8) is for existence of a solution a necessary condition. That is, if a solution exists, then (1.6.7) necessarily holds.

Thus let (1.6.1)

$$\lambda = \lambda_n = \frac{n^2\pi^2}{l^2}$$

and let u_n be an eigenfunction corresponding to that λ_n. Thus u_n satisfies

$$u_n'' + \lambda_n u_n = 0 \tag{1.6.9}$$

$$u_n(0) = 0 \tag{1.6.10}$$

$$u_n(l) = 0$$

Let the problem (1.6.1) and (1.6.2) have a solution u, that is, let

$$u'' + \lambda_n u = f \tag{1.6.11}$$

$$u(0) = 0 \tag{1.6.12}$$

$$u(l) = 0$$

be valid. Multiply (1.6.11) scalarly by the function u_n. We obtain

$$(u'', u_n) + \lambda_n (u, u_n) = (f, u_n) \tag{1.6.13}$$

However, in consequence of (1.6.10) and (1.6.12) we have

$$(u'', u_n) = \int_0^l u'' u_n \, dx = \left[u' u_n \right]_0^l - \int_0^l u' u_n' \, dx = - \int_0^l u' u_n' \, dx$$

$$= -\left[u u_n' \right]_0^l + \int_0^l u u_n'' \, dx = \int_0^l u u_n'' \, dx = (u, u_n'')$$

Thus by (1.6.13)

$$(f, u_n) = \left(u'', u_n\right) + \lambda_n \left(u, u_n\right) = \left(u, u_n''\right) + \lambda_n \left(u, u_n\right) = \left(u, u_n'' + \lambda_n u_n\right) = 0$$

since $u_n'' + \lambda_n u_n = 0$ by (1.6.9). Thus if for $\lambda = \lambda_n$ a solution of the problem in (1.6.1) and (1.6.2) exists, then necessarily $(f, u_n) = 0$ holds. According to Theorem 1.6.1, this condition is also sufficient for existence of a solution.

REMARK 1.6.1

As concerns only solvability of the problem (1.6.1) and (1.6.2) (thus not the task of finding the solution itself), which is often a question of essential importance (e.g., in stability problems), by Theorem 1.6.1 we can come to the desired conclusion on the basis of the corresponding homogeneous problem (1.6.3) and (1.6.4) only: If λ in (1.6.1) is not an eigenvalue of the problem (1.6.3) and (1.6.4), then the solvability questions "do not make difficulties"; if λ *is* an eigenvalue, then we verify whether the condition (1.6.7), that is, the condition (1.6.8), is satisfied. If it is, the problem (1.6.1) and (1.6.2) is solvable (and there is an infinite number of solutions); if it is not, then it is not solvable.

Example 1.6.1

Let us consider the problem

$$u'' - u = 1 \tag{1.6.14}$$

$$u(0) = 0 \tag{1.6.15}$$

$$u(1) = 0$$

This is a problem of the form

$$u'' + \lambda u = f \tag{1.6.16}$$

$$u(0) = 0 \tag{1.6.17}$$

$$u(l) = 0$$

with $l = 1$, $f(x) \equiv 1$, $\lambda = -1$. By Theorem 1.6.1, this problem is uniquely solvable, because $\lambda = -1$ is not an eigenvalue of the corresponding homogeneous problem

$$u'' + \lambda u = 0$$

$$u(0) = 0$$

$$u(1) = 0$$

(since this problem has only positive eigenvalues by Theorem 1.5.1).   ☐

Example 1.6.2
Let us decide on solvability of the problems

$$u'' + u = \sin x \tag{1.6.18}$$

$$u(0) = 0 \tag{1.6.19}$$

$$u(\pi) = 0$$

and

$$u'' + u = \cos x \tag{1.6.20}$$

$$u(0) = 0 \tag{1.6.21}$$

$$u(\pi) = 0$$

The problem (1.6.18) and (1.6.19) is a problem of the form

$$u'' + \lambda u = f$$

$$u(0) = 0$$

$$u(l) = 0$$

with $l = \pi$, $f(x) = \sin x$, and $\lambda = 1$. The corresponding homogeneous problem is

$$u'' + \lambda u = 0 \tag{1.6.22}$$

$$u(0) = 0 \tag{1.6.23}$$

$$u(\pi) = 0$$

and it has, by (1.5.25), the eigenvalues

$$\lambda_n = \frac{n^2\pi^2}{\pi^2} = n^2 \qquad (1.6.24)$$

$$n = 1, 2, \ldots$$

The corresponding eigenfunctions are

$$u_n = C_n \sin \frac{n\pi x}{\pi} = C_n \sin nx$$

$$C_n \neq 0 \text{ arbitrary}$$

Here, by (1.6.24), $\lambda = 1$ *is* an eigenvalue of (1.6.22) and (1.6.23). By Theorem 1.6.1, this problem is, or is not, solvable according to whether the orthogonality condition (1.6.8) is fulfilled. However, in our case the condition (1.6.8) reads

$$(\sin x, \sin x) = 0$$

However

$$(\sin x, \sin x) = \int_0^\pi \sin^2 x \, dx = \frac{\pi}{2} \neq 0$$

so that the problem (1.6.18) and (1.6.19) has no solution.

In the case of the problem (1.6.20) and (1.6.21), the situation is similar ($\lambda = 1$ is an eigenvalue of the corresponding homogeneous problem (1.6.22) and (1.6.23)), but the condition (1.6.8) is of the form

$$(\cos x, \sin x) = 0 \qquad (1.6.25)$$

Now

$$(\cos x, \sin x) = \int_0^\pi \cos x \sin x \, dx = \left[\frac{\sin^2 x}{2}\right]_0^\pi = 0$$

The condition in (1.6.25) is fulfilled, and the problem (1.6.20) and (1.6.21) is solvable (and has an infinite number of solutions). $\Box$

REMARK 1.6.2

Theorem 1.6.1 was applied to simple problems here, where we were able to verify their solvability even by direct computation. Usefulness of this theorem becomes clear only when the right-hand side of the given equation is "sufficiently complicated," so that the construction of the general solution is difficult, while the verification of the condition (1.6.8) is relatively easy. We have chosen the preceding simple examples so simple in order to be able to solve them directly and to compare the results with what we have concluded, using Theorem 1.6.1, in Examples 1.6.1 and 1.6.2; and to show the reader what happens in individual cases.

Let us start with the problem (1.6.14) and (1.6.15). Equation (1.6.14) evidently has a particular solution $u_p \equiv -1$. The general solution of the corresponding homogeneous equation $u'' - u = 0$ is $u_h = C_1 e^x + C_2 e^{-x}$ (K. Rektorys 1994, [1] Section 17.13). Thus the general solution of (1.6.14) is

$$u = -1 + C_1 e^x + C_2 e^{-x}$$

Boundary conditions in Equation (1.6.15) then yield

$$0 = -1 + C_1 + C_2$$

$$0 = -1 + C_1 e + C_2 e^{-1}$$

from which

$$C_1 = \frac{1}{1+e}$$

$$C_2 = \frac{e}{1+e}$$

and

$$u = -1 + \frac{e^x + e^{1-x}}{1+e}$$

The problem (1.6.14) and (1.6.15) is thus uniquely solvable, as stated in Example 1.6.1.

For the problem (1.6.18) and (1.6.19), the homogeneous differential equation

$$u'' + u = 0$$

corresponding to Equation (1.6.18), has the general solution

$$u_h = C_1 \cos x + C_2 \sin x$$

(K. Rektorys 1994, [1] Section 17.13). The particular solution u_p of (1.6.18) can be found in the form

$$u_p = x(A \cos x + B \sin x) \tag{1.6.26}$$

(K. Rektorys 1994, [1] Section 17.14). Hence

$$u'_p = A \cos x + B \sin x + x(-A \sin x + B \cos x) \qquad (1.6.27)$$

$$u''_p = 2(-A \sin x + B \cos x) - x(A \cos x + B \sin x)$$

Putting (1.6.26) and (1.6.27) into (1.6.18), we obtain

$$- 2A \sin x + 2B \cos x = \sin x \qquad (1.6.28)$$

from which, comparing coefficients, we get

$$A = -\frac{1}{2}$$

$$B = 0$$

Thus the general solution of (1.6.18) is

$$u = u_p + u_h = -\frac{1}{2}x \cos x + C_1 \cos x + C_2 \sin x$$

The first condition (1.6.19) implies $C_1 = 0$, so that

$$u = -\frac{1}{2}x \cos x + C_2 \sin x$$

The second one then yields

$$0 = -\frac{1}{2}\pi \cos \pi + C_2 \sin \pi \qquad (1.6.29)$$

However, $\cos \pi = -1$, $\sin \pi = 0$, so that the condition (1.6.29) cannot be fulfilled by any choice of C_2. Consequently, the problem (1.6.18) and (1.6.19) is not solvable, as we have stated in Example 1.6.2 (see also Problem 1.10.4).

For the problem (1.6.20) and (1.6.21), the situation is similar; the only difference is that on the right-hand side of (1.6.28) we have $\cos x$ instead of $\sin x$. Consequently, $A = 0$, $B = \frac{1}{2}$, and the general solution of (1.6.20) is

$$u = \frac{1}{2}x \sin x + C_1 \cos x + C_2 \sin x$$

The first condition in (1.6.21) gives $C_1 = 0$, thus

$$u = \frac{1}{2}x \sin x + C_2 \sin x$$

However, the second condition (1.6.21) is fulfilled for any C_2, because of $\sin \pi = 0$. Thus the problem (1.6.20) and (1.6.21) has an infinite number of solutions, as we have affirmed in Example 1.6.2.

REMARK 1.6.3   *(generalization of preceding results)*

1. In Theorem 1.6.1, we considered the interval $[0, l]$. A similar theorem can be given for an arbitrary (bounded) interval $[a, b]$.

2. Further, we assumed that the right-hand side f of the given equation is continuous in the interval $[0, l]$. If we modify suitably the concept of a solution, Theorem 1.6.1 remains valid under the mere assumption that $f \in L_2(0, l)$. We do not go into detail here, since in Section 3.6 we arrive at essentially more general results. One of their consequences is Theorem 1.6.2 that follows, often useful in applications.

Consider the problem

$$u'' + \lambda gu = f \tag{1.6.30}$$

$$u(0) = 0 \tag{1.6.31}$$

$$u(l) = 0$$

and the corresponding homogeneous problem

$$u'' + \lambda gu = 0 \tag{1.6.32}$$

$$u(0) = 0 \tag{1.6.33}$$

$$u(l) = 0$$

Here f and g are continuous functions in the interval $[0, l]$; g is positive in that interval. (Especially for $g \equiv 1$ we get the previous problems (1.6.1), (1.6.2), (1.6.3), and (1.6.4).) Then we have the theorem that follows.

THEOREM 1.6.2
The problem (1.6.32) and (1.6.33) has a countable set

$$\lambda_1, \lambda_2, \ldots$$

of eigenvalues, all of them being positive.

1. Let, in (1.6.30), λ be not equal to any of them. Then the problem (1.6.30) and (1.6.31) is uniquely solvable for every right-hand side f (continuous in $[0, l]$).

2. If λ is equal to an eigenvalue λ_n, then the problem (1.6.30) and (1.6.31) has no solution, in general. It is solvable (and in this case it has an infinite number of solutions) exactly if the function f is orthogonal, in $L_2(0, l)$, to every eigenfunction corresponding to that eigenvalue.

See also Problem 1.10.13.

REMARK 1.6.4
Especially under the assumption of positivity of the function g in $[0, l]$, the problem

$$u'' - gu = f \tag{1.6.34}$$

$$u(0) = 0 \tag{1.6.35}$$

$$u(l) = 0$$

is uniquely solvable for every (continuous) right-hand side f (because $\lambda = -1$ cannot be an eigenvalue of the problem (1.6.32) and (1.6.33), since that problem has, by Theorem 1.6.2, only positive eigenvalues; cf. Example 1.6.1).

REMARK 1.6.5
Let us note that, in contrast to the preceding examples, Equation (1.6.30) or (1.6.32) is not an equation with constant coefficients, where a characteristic (= auxiliary) equation can be applied to construct the general solution, and that the eigenvalues of (1.6.32) and (1.6.33) as well as solutions of (1.6.30) and (1.6.31) can be exactly determined only in simplest cases. However, just these problems are often encountered in applications (deflection, or buckling of bars with variable cross sections, etc.). Therefore, in the following text we present a very simple method of approximate solution to these problems—the finite-difference method, or the net method.

1.7 The Finite-Difference Method for Ordinary Differential Equations

The *finite-difference method* (or the *net method*) consists of replacing derivatives in the given differential equation by corresponding difference quotients.

Remember that the derivative of a function u at the point x_0 is defined by the limit

$$u'(x_0) = \lim_{h \to 0} \frac{u(x_0 + h) - u(x_0)}{h} = \lim_{h \to 0} \frac{u_1 - u_0}{h}$$

where the notation in Figure 1.7.1 has been used. If h is "small," then it is possible to replace the derivative at point x_0 by the difference quotient $(u_1 - u_0)/h$. We write

$$u'(x_0) \approx \frac{u_1 - u_0}{h} \tag{1.7.1}$$

The second derivative at point x_0 is defined, as we know, as a derivative of the first derivative. Thus we replace it by the second difference quotient, which is the difference quotient of the first difference quotient

$$u''(x_0) \approx \frac{\frac{u_1 - u_0}{h} - \frac{u_0 - u_2}{h}}{h}$$

(for the notation see Figure 1.7.1). Thus

$$u''(x_0) \approx \frac{u_1 - 2u_0 + u_2}{h^2} \tag{1.7.2}$$

The second derivative of the function u at the given point is thus replaced by a difference quotient in the denominator, of which we have h^2; and in the nominator we have the value of u at that point, multiplied by the coefficient -2, and the values of u at "neighboring" points with the coefficient $+1$.

There are other difference approximations, if wanted. See, for example, K. Rektorys 1994, [1] Chapter 27, where also difference approximations for derivatives of higher orders can be found, inclusive of orders of accuracy (cf. also Problem 1.10.8). Next we shall use approximations (1.7.1) and (1.7.2).

How to approximately solve a given problem by the method of finite differences will be shown in the following example.

Example 1.7.1
Let us solve the problem

$$u'' - (1 + \sin^2 x)u = -4 \tag{1.7.3}$$

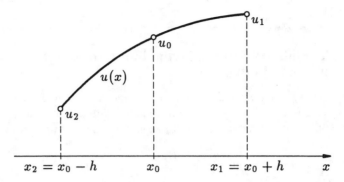

FIGURE 1.7.1

$$u(0) = 0 \qquad\qquad (1.7.4)$$

$$u(\pi) = 0$$

approximately by the finite-difference method with the step $h = \pi/3$.

First, note that the given problem is uniquely solvable. (See Remark 1.6.4; the functions $f(x) \equiv -4$ and $g(x) = 1 + \sin^2 x$ are continuous in the interval $[0, \pi]$, with $g(x)$ being a positive function.)

Let us divide the interval $[0, \pi]$ into three subintervals of the length $h = \pi/3$ by points $x_1 = \pi/3$ and $x_2 = 2\pi/3$. Denote by z_i, $i = 0, 1, 2, 3$, the wanted values of the so-called *finite-difference solution* (*finite difference approximation*, or *net solution*, in brief), obtained by the finite-difference method at points x_i (see Figure 1.7.2). We have to realize that these values will be different from the values

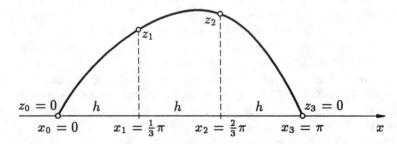

FIGURE 1.7.2

$u_i = u(x_i)$ of the exact solution at these points, in general, because our method is an approximate one. At points $x_0 = 0$ and $x_3 = \pi$ we put, of course, $z_0 = 0$ and $z_3 = 0$ by (1.7.4). We thus have to compute z_1 and z_2. To this aim, replace the given equation at points x_1 and x_2 by corresponding difference equations that arise when replacing the derivatives at these points by corresponding difference quotients. The values of the function $1 + \sin^2 x$ are to be evaluated at points x_1 and x_2 separately, in general. Because we have

$$\sin \frac{\pi}{3} = \frac{\sqrt{3}}{2}$$

$$\sin \frac{2\pi}{3} = \frac{\sqrt{3}}{2}$$

here, both these values are equal to $1 + \frac{3}{4}$. The difference equation at the point x_1 thus becomes

$$\frac{z_2 - 2z_1 + z_0}{\left(\frac{1}{3}\pi\right)^2} - \left(1 + \frac{3}{4}\right) z_1 = -4 \tag{1.7.5}$$

at the point x_2

$$\frac{z_3 - 2z_2 + z_1}{\left(\frac{1}{3}\pi\right)^2} - \left(1 + \frac{3}{4}\right) z_2 = -4 \tag{1.7.6}$$

This is a system of two equations for two unknowns z_1 and z_2 (the values $z_0 = 0$ and $z_3 = 0$ being known). To solve such a system is easy. In our case it is particularly easy, because by symmetry (caused by the same values of the function $1 + \sin^2 x$ and of the right-hand side at points x_1 and x_2 here) it follows that

$$z_2 = z_1$$

Let us prove it: Subtract Equation (1.7.5) from Equation (1.7.6). We obtain (with $z_0 = 0$ and $z_3 = 0$ already set)

$$\frac{-3 (z_2 - z_1)}{\frac{1}{9}\pi^2} - \frac{7}{4} (z_2 - z_1) = 0$$

that is

$$\left(\frac{27}{\pi^2} + \frac{7}{4}\right) (z_1 - z_2) = 0$$

This yields immediately $z_2 = z_1$. Putting this result into (1.7.5) we obtain

$$-\frac{z_1}{\frac{1}{9}\pi^2} - \frac{7}{4}z_1 = -4$$

from which

$$z_1 \doteq 1.503$$

and, consequently

$$z_2 \doteq 1.503$$

See also Problems 1.10.1 to 1.10.3.   []

1.8   Convergence of the Finite-Difference Method

As can be easily seen from the presented example, the finite-difference method gives approximations of the solution at certain points of the given interval only (in our case at points x_1 and x_2).

To obtain an approximation in the whole interval, construct a piecewise linear function—denote it by w_1—which assumes, at the endpoints and at the points of division of the considered interval, values of the finite-difference solution (Figure 1.8.1). A natural question arises as to how "close" we are to the exact solution with our approximation, in other words, how much the function w_1 differs from the solution u. Intuitively it can be expected that this difference will be smaller, the finer the division of the given interval. In this way we come to the problematics of *convergence of the finite-difference method*. Let us formulate the corresponding result for the problem in Remark 1.6.4

$$u'' - gu = f \tag{1.8.1}$$

$$u(0) = 0 \tag{1.8.2}$$

$$u(l) = 0$$

where f and g are continuous functions in the interval $[0, l]$ and $g(x) > 0$ for all x of that interval. Choose a basic division of the interval $[0, l]$ (e.g., into three subintervals of the same lengths, as we did in Example 1.7.1), obtain the finite-difference solution, and construct the piecewise linear function w_1 as indicated earlier. Then let us construct another division, bisecting each of subintervals of the preceding division, and construct the piecewise linear function w_2 as we previously had constructed the function w_1. If we go on in this way, we obtain a sequence of functions w_n.

It is possible to show that if functions f and g are sufficiently smooth (we do not go into detail; in Example 1.7.1 these functions have even derivatives of all orders), the sequence $\{w_n\}$ converges uniformly to the exact solution u of the problem (1.8.1) and (1.8.2). See, for example, K. Rektorys 1994, [1] Chapter 27, where information about an error estimate also can be found. Let us note that,

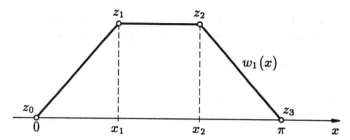

FIGURE 1.8.1

even though the idea of the finite-difference method itself is very simple, questions connected with error estimates are complicated.

All that has been said about the problem (1.8.1) and (1.8.2) considered on the interval $[0, l]$ remains valid for an arbitrary (bounded) interval $[a, b]$. Finally, let us note that for the finite-difference method it is not essential whether at the end-points of the given interval only zero values for the solution are prescribed.

1.9 Application of the Finite-Difference Method in Eigenvalue Problems

The finite-difference method can be applied also to the approximate solution of eigenvalue problems. Let us present an example.

Example 1.9.1
Find approximately the first (i.e., the smallest) two eigenvalues of the problem

$$u'' + \lambda(4 + \sin^2 x)u = 0 \tag{1.9.1}$$

$$u(0) = 0 \tag{1.9.2}$$

$$u(\pi) = 0$$

This is a problem of the type (1.6.32) and (1.6.33), with $l = \pi$ and $g = 4 + \sin^2 x$. (The given equation thus has variable coefficients.)

Let us divide the interval $[0, \pi]$, as in Example 1.7.1, into three subintervals by points $x_1 = \pi/3$ and $x_2 = 2\pi/3$. Denote by μ approximation of the eigenvalue λ obtained by the finite-difference method, and by z_1 and z_2 corresponding values of the finite-difference solution at points x_1 and x_2. Because the finite-difference solution has to be an approximation of the eigenfunction u, we require that it be a nonzero one, that is, that at least one of the numbers z_1, z_2 be different from zero. At points $x_0 = 0$ and $x_3 = \pi$ we put, of course, $z_0 = 0$ and $z_3 = 0$ by (1.9.2). Now, if we write at points x_1 and x_2 Equation (1.9.1) in the difference form, we obtain (similarly to Example 1.7.1)

$$\frac{z_2 - 2z_1 + z_0}{\frac{1}{9}\pi^2} + \mu\left(4 + \frac{3}{4}\right)z_1 = 0$$

$$\frac{z_3 - 2z_2 + z_1}{\frac{1}{9}\pi^2} + \mu\left(4 + \frac{3}{4}\right)z_2 = 0$$

that is, (because $z_0 = 0$ and $z_3 = 0$)

$$\left(\frac{19}{4}\mu - \frac{18}{\pi^2}\right)z_1 + \frac{9}{\pi^2}z_2 = 0 \qquad (1.9.3)$$

$$\frac{9}{\pi^2}z_1 + \left(\frac{19}{4}\mu - \frac{18}{\pi^2}\right)z_2 = 0 \qquad (1.9.4)$$

As well known from linear algebra, the homogeneous system (1.9.3) and (1.9.4) has a nonzero solution exactly if its determinant is equal to zero, that is, exactly if

$$\left(\frac{19}{4}\mu - \frac{18}{\pi^2}\right)^2 - \frac{81}{\pi^4} = 0 \qquad (1.9.5)$$

This is thus the condition for μ guaranteeing that the finite-difference solution, approximating the eigenfunction, not be a zero one. If we denote by μ_1 and μ_2 ($\mu_1 < \mu_2$) roots of quadratic equation (1.9.5), then (writing that equation in the form

$$\left(\frac{19}{4}\mu - \frac{18}{\pi^2}\right)^2 = \frac{81}{\pi^4}$$

from which

$$\frac{19}{4}\mu - \frac{18}{\pi^2} = \pm\frac{9}{\pi^2}\Bigg)$$

we obtain

$$\frac{19}{4}\mu_1 = -\frac{9}{\pi^2} + \frac{18}{\pi^2}$$

$$\frac{19}{4}\mu_2 = +\frac{9}{\pi^2} + \frac{18}{\pi^2}$$

Consequently

$$\mu_1 = \frac{4}{19} \cdot \frac{9}{\pi^2} \doteq 0.192 \qquad (1.9.6)$$

$$\mu_2 = \frac{4}{19} \cdot \frac{27}{\pi^2} \doteq 0.576 \qquad (1.9.7)$$

It can be shown that μ_1 and μ_2 are just approximations of the first (= least) eigenvalues λ_1 and λ_2 of the problem considered, while the approximation μ_1 of λ_1 is substantially better than that of λ_2 by μ_2 (which the reader may easily verify in the case of problems with constant coefficients). This is of use when solving problems arising in civil engineering, for example, because just the first eigenvalue λ_1 is most interesting, as usual. See also Problem 1.10.10. ☐

1.10 Problems 1.10.1 to 1.10.15

1.10.1—By using the finite-difference method with step $h = \frac{2}{3}$, solve the problem

$$u'' - \frac{1}{4}\left(1 + \sin^2 \pi x\right) u = 1 - x \qquad (1.10.1)$$

$$u(-1) = 0 \qquad (1.10.2)$$

$$u(1) = 0$$

Make a "check of reasonability" of the obtained approximate values z_1 and z_2 at points $x_1 = -\frac{1}{3}$ and $x_2 = \frac{1}{3}$, comparing them with values v_1 and v_2 of exact solution v of the "simplified problem"

$$v'' = 1 - x \qquad (1.10.3)$$

$$v(-1) = 0 \qquad (1.10.4)$$

$$v(1) = 0$$

at these points.

$$[v = -\frac{x^3}{6} + \frac{x^2}{2} + \frac{x}{6} - \frac{1}{2}$$

$$z_1 \doteq -0.418 \tag{1.10.5}$$

$$z_2 \doteq -0.326$$

$$v_1 \doteq -0.494 \tag{1.10.6}$$

$$v_2 \doteq -0.395 \ ]$$

1.10.2—For the problem

$$u'' - (1 + \sin^2 x)u = -4 \tag{1.10.7}$$

$$u(0) = 0 \tag{1.10.8}$$

$$u(\pi) = 0$$

from Example 1.7.1 and for the "simplified" problem

$$v'' = -4 \tag{1.10.9}$$

$$v(0) = 0 \tag{1.10.10}$$

$$v(\pi) = 0$$

we get, similar to the way in Problem 1.10.1, with $h = \frac{\pi}{3}$, $x_1 = \frac{\pi}{3}$, $x_2 = \frac{2\pi}{3}$, $v_1 = v(x_1)$ and $v_2 = v(x_2)$

$$z_1 \doteq 1.503 \tag{1.10.11}$$

$$z_2 \doteq 1.503$$

$$v_1 \doteq 4.387 \tag{1.10.12}$$

$$v_2 \doteq 4.387$$

In each of Problems 1.10.1 and 1.10.2 there is a certain difference between the values z_1 and z_2 and v_1 and v_2. Something like this could be expected, because

1. In both problems only two nodes (= division points) x_1 and x_2 have been chosen; thus the finite-difference method gives rough approximations only.

2. Equations (1.10.1) and (1.10.3), or Equations (1.10.7) and (1.10.9), respectively, are different.

In Problem 1.10.1 the difference between z_1 and v_1, or z_2 and v_2 is relatively small (does not exceed 20%), and (1.10.6) may be considered as a confirmation of (1.10.5) in a certain sense. In Problem 1.10.2 the difference (in percentage) between (1.10.11) and (1.10.12) is much more remarkable. Why?

1.10.3—By the finite-difference method with step $h = \frac{1}{3}$ solve the problem

$$u'' - xu = -2 \tag{1.10.13}$$

$$u(0) = 0 \tag{1.10.14}$$

$$u(1) = 0$$

[$z_1 \doteq 0.212$, $z_2 \doteq 0.209$; let us note that function $g(x) = x$ does not satisfy the assumptions stated on page 35 exactly, not being positive in the whole interval $[0, 1]$ (at point $x = 0$ it is equal to zero). However, it will be shown in Chapter 3 that even nonnegativity of function $g(x)$ is sufficient for existence and uniqueness of solution of the given problem.]

1.10.4—By using the finite-difference method with step $h = \pi/4$, solve the problem

$$u'' + u = \sin x$$

$$u(0) = 0$$

$$u(\pi) = 0$$

[To solve this problem by any approximate method makes no sense, because no solution exists—see Example 1.6.2. Can something similar happen in the preceding problems?]

1.10.5—(Nonhomogeneous boundary conditions)

1. By the finite-difference method with step $h = \frac{1}{2}$ solve the problem

$$u'' - (1 + x^2)u = 2$$

$$u(0) = 2$$

$$u(2) = -1$$

2. By using the substitution

$$u = v - \frac{3}{2}x + 2$$

convert the problem into one with homogeneous boundary conditions and solve it by the same method (Figure 1.10.1).

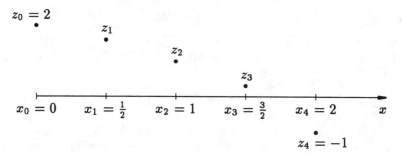

FIGURE 1.10.1

$$[\; z_1 \doteq 0.551$$

$$z_2 \doteq -0.225$$

$$z_3 \doteq -0.613$$

$$v'' - (1 + x^2)v = 4 - \frac{3}{2}x + 2x^2 - \frac{3}{2}x^3$$

$$v(0) = 0$$

$$v(2) = 0$$

$$v_1 \doteq -0.699$$

$$v_2 \doteq -0.725$$

$$v_3 \doteq -0.363$$

Check:

$$u_1 = v_1 - \frac{3}{2} \cdot \frac{1}{2} + 2 = \quad 0.551 = z_1$$

$$u_2 = v_2 - \frac{3}{2} \cdot 1 + 2 = -0.225 = z_2$$

$$u_3 = v_3 - \frac{3}{2} \cdot \frac{3}{2} + 2 = -0.613 = z_3 \ ]$$

1.10.6—(Derivative(s) in boundary conditions) In Problem 3.7.8 we shall be interested in solving the (sample) problem

$$u'' = -1 \tag{1.10.15}$$

$$u(0) = 0, \quad u'(1) = 0 \tag{1.10.16}$$

by the classical Ritz method as well as by the finite-element method. Here the reader is asked to solve this problem

1. Exactly

2. By the finite-difference method with step $h = \frac{1}{4}$, to be able then to compare numerical results obtained by all these methods at the points $x_1 = \frac{1}{4}$, $x_2 = \frac{2}{4}$, $x_3 = \frac{3}{4}$.

Hint:

1. $u = x - \frac{1}{2}x^2$; $u(\frac{1}{4}) = \frac{7}{32}$, $u(\frac{2}{4}) = \frac{12}{32}$, and $u(\frac{3}{4}) = \frac{15}{32}$

2. Choose $z_4 = z_3$ to take the second condition of (1.10.16) into account (Figure 1.10.2); we get $z_1 = \frac{6}{32}$, $z_2 = \frac{10}{32}$, and $z_3 = z_4 = \frac{12}{32}$.

1.10.7—Find the deflection of a bar loaded according to Figure 1.10.3 (cf. Figure 1.1.1).

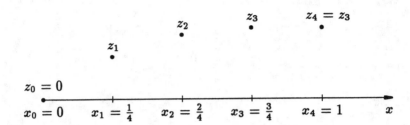

FIGURE 1.10.2

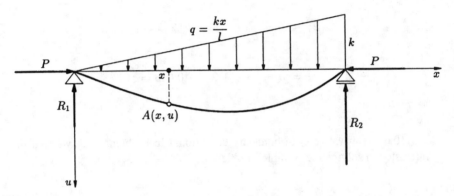

FIGURE 1.10.3

Hint: First we have to find the reactions R_1 and R_2; Equation of

1. Force equilibrium

$$R_1 + R_2 = \int_0^l \frac{kx}{l}dx = \frac{kl}{2}$$

2. Moment equilibrium (with respect to the point $(0, 0)$)

$$l R_2 = \int_0^l x \cdot \frac{kx}{l}dx = \frac{kl^2}{3}$$

Thus

$$R_2 = \frac{kl}{3}$$

$$R_1 = \frac{kl}{6}$$

Further (cf. Section 1.1)

$$M = R_1 x + Pu - \int_0^x (x-t)\frac{kt}{l}dt = \frac{kl}{6}x + Pu - \frac{k}{6l}x^3$$

$$u'' + a^2 u = bx^3 + cx$$

with

$$a^2 = \frac{P}{EI}$$

$$b = \frac{k}{6EIl}$$

$$c = -\frac{kl}{6EI}$$

$$u(0) = 0$$

$$u(l) = 0$$

$$u = \frac{b}{a^2}x^3 + \left(\frac{c}{a^2} - \frac{6b}{a^4}\right)x + \frac{6bl}{a^4 \sin al}\sin ax$$

provided $\sin al \neq 0$; the critical force remains the same as in Section 1.1.

1.10.8—Consider the problem

$$u'''' = 1 \tag{1.10.17}$$

$$u(0) = 0, \quad u(1) = 0 \tag{1.10.18}$$

$$u'(0) = 0, \quad u'(1) = 0 \tag{1.10.19}$$

(This is deflection of a clamped uniformly loaded bar.) In this very simple case, the explicit solution can be easily found to be

$$u = \frac{1}{24}x^2(1-x)^2 \tag{1.10.20}$$

(How?) Nevertheless, solve the problem by the finite-difference method with step $h = \frac{1}{3}$, replacing, as usual, u'''' at points x_1 or x_2 (see Figure 1.10.4) by

$$u''''(x_1) \approx \frac{z_3 - 4z_2 + 6z_1 - 4z_0 + z_{-1}}{h^4}$$

or

$$u''''(x_2) \approx \frac{z_4 - 4z_3 + 6z_2 - 4z_1 + z_0}{h^4}$$

respectively, and taking the conditions in (1.10.19) into account using $z_{-1} = z_0$ and $z_4 = z_3$. Compare the obtained values z_1 and z_2 with values of the exact solution (1.10.20) at points x_1 and x_2.

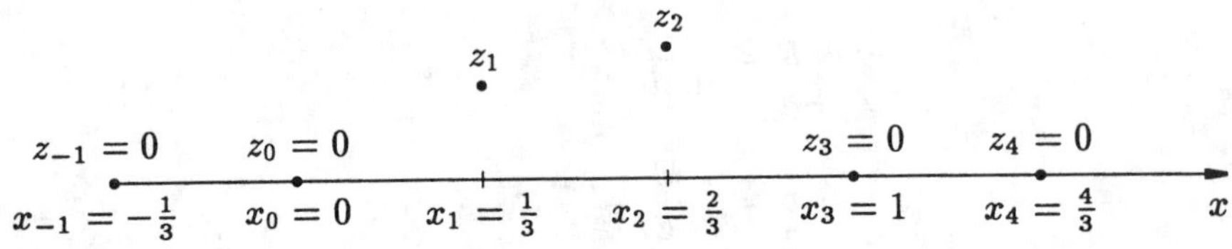

FIGURE 1.10.4

[$u(x_1) = u(x_2) = \frac{1}{486}$; $z_1 = z_2 = \frac{1}{162}$. What is the reason for such a difference? (Solve the same problem with a finer h.)]

1.10.9—By Theorem 1.5.2, the system of functions

$$\sin \frac{n\pi x}{l}$$

$$n = 1, 2, \ldots$$

is orthogonal in $L_2(0, l)$. Make yourself sure of it by direct computation.

Hint: Use the well-known formula

$$\sin \alpha \sin \beta = \frac{1}{2}\{\cos(\alpha - \beta) - \cos(\alpha + \beta)\}$$

1.10.10—By using the finite-difference method with step $h = 1$, find approximations μ_1 and μ_2 of the first two eigenvalues λ_1 and λ_2 of the problem

$$u'' + \lambda(3 + x)u = 0$$

$$u(0) = 0$$

$$u(3) = 0$$

[$\mu_1 \doteq 0.221$ and $\mu_2 \doteq 0.679.$]

1.10.11*—By following the proofs of Theorems 1.5.1 and 1.5.2, prove that the problem

$$u'' + \lambda u = 0 \tag{1.10.21}$$

$$u(0) = 0 \tag{1.10.22}$$

$$u'(l) = 0$$

1. May have positive eigenvalues only.

2. Eigenfunctions u_i and u_j corresponding to different eigenvalues λ_i and λ_j are orthogonal in $L_2(0, l)$.

1.10.12—By using (1) from the preceding problem, show that eigenvalues, or eigenfunctions of problem (1.10.21) and (1.10.22) are

$$\lambda_n = \frac{(2n - 1)^2 \pi^2}{4l^2} \tag{1.10.23}$$

$$n = 1, 2, \ldots$$

or

$$u_n = C_n \sin \frac{(2n - 1)\pi x}{2l} \tag{1.10.24}$$

$$n = 1, 2, \ldots$$

$$C_n \neq 0 \text{ arbitrary}$$

respectively. Establish orthogonality of the functions u_i, u_j, $i \neq j$, by direct computation.

Hint: Cf. Problem 1.10.9.

1.10.13*—Consider the problems

$$u'' + \lambda g u = f \tag{1.10.25}$$

$$u(0) = 0 \tag{1.10.26}$$

$$u(l) = 0$$

and

$$u'' + \lambda g u = 0 \tag{1.10.27}$$

$$u(0) = 0 \tag{1.10.28}$$

$$u(l) = 0$$

from page 35. Prove that under the assumption

$$f(x) \text{ and } g(x) \text{ are continuous functions in } [0, l] \tag{1.10.29}$$

$$g(x) > 0$$

1. The problem (1.10.27) and (1.10.28) may have positive eigenvalues only.

2. Eigenfunctions u_i and u_j corresponding to different eigenvalues λ_i and λ_j are orthogonal in $L_2(0, l)$ with the "weight" $g(x)$; that is, we have

$$\int_0^l g(x) u_i(x) u_j(x) dx = 0$$

3. If λ_i is an eigenvalue of the problem (1.10.27) and (1.10.28), then the necessary condition for the problem (1.10.25) and (1.10.26) to be solvable is

$$(f, u_i) = 0$$

where u_i is any eigenfunction corresponding to the eigenvalue λ_i.

Hint: Follow Section 1.5 and the proof of orthogonality of f and u_i in Section 1.6.

1.10.14* — Consider the eigenvalue problem

$$u'''' - \lambda u = 0 \tag{1.10.30}$$

$$u(0) = 0 \tag{1.10.31}$$

$$u(l) = 0$$

$$u'(0) = 0$$

$$u'(l) = 0$$

Show that

1. The eigenvalues λ_n may be positive only.

2. Put $\lambda_n = k_n^4$, $k_n > 0$. Then the values of k_n are roots of the equation

$$\cos kl \cdot \cosh kl = 1 \tag{1.10.32}$$

Hint: For (1) follow Section 1.5.

3. For 2, write the general integral of the equation

$$u'''' - k^4 u = 0$$

and use conditions (1.10.31) to determine C_1, C_2, C_3 and C_4. You obtain a homogeneous system of linear equations. $C_1, \ldots, C_4$ must not vanish simultaneously, because the solution should be an eigenfunction. This implies that the determinant of the system is zero, which leads, after rearranging, to the condition (1.10.32).

1.10.15*—An equation of the form

$$\big(a(x)u'\big)' + c(x)u = f(x) \tag{1.10.33}$$

is said to be written in a *selfadjoint* (or *divergent*) form. Show that any equation of the form

$$A(x)u'' + B(x)u' + C(x)u = F(x) \tag{1.10.34}$$

with B, C, F continuous, A continuously differentiable, and $A(x) > 0$ in an interval J, can be converted into a selfadjoint form in that interval. Apply this result to the equation

$$xu'' + \big(1 + x^2\big)u' + (3 - x)u = F(x) \tag{1.10.35}$$

Hint: Multiply (1.10.34) by an unknown function $k(x)$ and denote $a(x) = k(x)A(x)$. Differentiation

$$(au')' = au'' + a'u'$$

gives a condition for k—it should be a solution of the linear first-order equation

$$k' + \frac{A' - B}{A} k = 0$$

For Equation (1.10.35) we get

$$k = e^{x^2/2}$$

Chapter 2

Partial Differential Equations: Classical Approach

2.1 Basic Concepts; Examples of Equations Frequently Encountered in Applications; The Heat-Conduction Equation Derived

A differential equation is called *partial*, if the function to be found is a function of two or more variables.

Thus such an equation contains partial derivatives of that function. The general form of a partial differential equation is

$$F\left(x_1, \ldots, x_N, u, \frac{\partial u}{\partial x_1}, \ldots, \frac{\partial u}{\partial x_N}, \frac{\partial^2 u}{\partial x_1^2}, \ldots, \frac{\partial^k u}{\partial x_1^k}, \ldots\right) = 0 \qquad (2.1.1)$$

The order of the highest derivative appearing in this equation is called the *order* of the equation.

Under a *solution* of Equation (2.1.1) in a domain $\Omega \subset E_N$ we understand such a (sufficiently smooth) function $u(x_1, \ldots, x_N)$ that if substituted, including its derivatives, into Equation (2.1.1), satisfies that equation identically (i.e., everywhere) in Ω.

Example 2.1.1

Consider in $\Omega = E_2$ (thus in the xy-plane, if we denote the variables by x, y instead of x_1, x_2, as usual) the equation

$$\frac{\partial^2 u}{\partial x^2} + \frac{\partial^2 u}{\partial y^2} = 0 \qquad (2.1.2)$$

This is a partial differential equation of the second order for the function $u(x, y)$.

An example of a solution of Equation (2.1.2) in $\Omega = E_2$ is the function

$$u = x^2 - y^2 \tag{2.1.3}$$

since

$$\frac{\partial^2 u}{\partial x^2} = 2$$

$$\frac{\partial^2 u}{\partial y^2} = -2$$

at every point $(x, y) \in E_2$, so that Equation (2.1.2) is fulfilled everywhere in E_2 for this function. Similarly, it is possible to verify that the functions

$$u = e^y \sin x \tag{2.1.4}$$

$$u = \sin x \cosh y$$

(and many others) are also solutions of Equation (2.1.2) in $\Omega = E_2$. $\quad\square$

Equation (2.1.2), very often encountered in applications, is the so-called *Laplace equation*. (P. S. Laplace was an outstanding French mathematician, also well known in physics and astronomy. He devoted much effort to the study of this equation.) It is usual to write this equation briefly in the form

$$\Delta u = 0 \tag{2.1.5}$$

where

$$\Delta = \frac{\partial^2}{\partial x^2} + \frac{\partial^2}{\partial y^2}$$

is the so-called *Laplace operator*. In E_N this operator is of the form

$$\Delta = \frac{\partial^2}{\partial x_1^2} + \cdots + \frac{\partial^2}{\partial x_N^2} \tag{2.1.6}$$

To solve Equation (2.1.1) means—as in the case of ordinary differential equations—to find all its solutions. As we know, in the case of an ordinary differential equation it is possible to find its so-called *general solution,* which contains (at least in a certain region) all its solutions. In the case of partial differential

equations something like this is not possible, in general. In the case of partial
equations of very special types we are able to find a general form of their solutions.
Such a "general solution" then depends on one, or more "arbitrary" functions. An
example of such an equation is

$$\frac{\partial^2 u}{\partial x \partial y} = 0 \tag{2.1.7}$$

Integrating with respect to y we obtain

$$\frac{\partial u}{\partial x} = f(x)$$

and, then, integrating with respect to x

$$u = \int f(x)\, dx + h(y) = g(x) + h(y) \tag{2.1.8}$$

where g is a primitive function to f. The function $u(x, y) = g(x) + h(y)$ is
obviously a solution of Equation (2.1.7) for arbitrary (differentiable) functions g
and h; conversely, the way in which we have derived this result shows that every
solution of Equation (2.1.7) is contained in (2.1.8).

We are not going into detail about this problematic, because we are interested
in other questions. How to effectively solve partial differential equations—in
particular, with so-called boundary conditions—will be shown in the next chapters.
Here we first present, typical equations encountered in applications, especially the
Poisson and Laplace equations, the plate equation, the biharmonic equation, the
heat-conduction equation, and the wave equation.

1. The *Poisson equation*

$$\Delta u = f \tag{2.1.9}$$

where Δ is the Laplace operator (see (2.1.6)) and f is a given function of
$x_1, \ldots, x_N$. For $N = 2$ the equation becomes (writing, as usual, x, y instead
of x_1, x_2)

$$\frac{\partial^2 u}{\partial x^2} + \frac{\partial^2 u}{\partial y^2} = f(x, y) \tag{2.1.10}$$

2. A special case of the Poisson equation is the Laplace equation

$$\Delta u = 0 \tag{2.1.11}$$

especially for $N = 2$, the equation

$$\frac{\partial^2 u}{\partial x^2} + \frac{\partial^2 u}{\partial y^2} = 0 \tag{2.1.12}$$

3. The equation

$$\Delta^2 u = f \tag{2.1.13}$$

where (we consider only the plane case, i.e., $N = 2$)

$$\Delta^2 u = \frac{\partial^4 u}{\partial x^4} + 2\frac{\partial^4 u}{\partial x^2 \partial y^2} + \frac{\partial^4 u}{\partial y^4}$$

is called the *equation of deflection of a plate*, or the *plate equation*, in brief.

4. Its special case is the *biharmonic equation*

$$\Delta^2 u = 0$$

5. The equation

$$\frac{\partial u}{\partial t} = a^2 \Delta u + f \tag{2.1.14}$$

is the so-called *heat-conduction equation* (in chemical problems the *diffusion equation*). In applications, we assume $N \overset{\leq}{=} 3$, of course. In heat-conduction problems, $u(x_1, \ldots, x_N, t)$ is the temperature, t is time

$$a^2 = \frac{\lambda}{\rho c}$$

is the so-called *diffusivity*; here λ is conductivity, ρ is specific mass, and c is specific heat of the material. The function f characterizes intensity of inner heat sources. (*Note:* In a conductor, heat is developed if electrical current is going through; in stiffening concrete, heat is developed because of a chemical reaction running through in the cement.) The simplest special case of Equation (2.1.14) is *one-dimensional* heat-conduction equation, without inner heat sources

$$\frac{\partial u}{\partial t} = a^2 \frac{\partial^2 u}{\partial x^2} \tag{2.1.15}$$

6. The *wave equation* (*equation of oscillations*) is

$$\frac{\partial^2 u}{\partial t^2} = a^2 \Delta u \tag{2.1.16}$$

Its special case is the *equation of a vibrating string* (= of a perfectly flexible fiber)

$$\frac{\partial^2 u}{\partial t^2} = a^2 \frac{\partial^2 u}{\partial x^2} \tag{2.1.17}$$

where $a^2 = F/\rho$; ρ is the mass of the string per unit length and F is the magnitude of the tensile force F. To derive Equation (2.1.17) see Problem 2.4.4.

REMARK 2.1.1   *(informational note).*

Equations mentioned in (1), (2), (3), and (4) belong to the so-called *elliptical type* of partial differential equations; Equations (2.1.14) and (2.1.15) are examples of *parabolic equations* and Equations (2.1.16) and (2.1.17) are examples of *hyperbolic equations.* Very roughly speaking, elliptical equations describe stationary phenomena, thus not depending on time; parabolic equations describe problems in heat conduction or diffusion, and hyperbolic equations describe those in oscillations, vibration, and so forth. Exact definitions can be found in K. Rektorys 1994, [1] Section 18.3.

REMARK 2.1.2

Without exaggerating, it is possible to say that partial differential equations govern the whole of theoretical physics, or theoretical fields that separated from it as independent disciplines (theory of elasticity, dynamics, etc.). To derive a differential equation, describing a certain physical process, is not easy, in general. Let us establish here, for illustration, the one-dimensional heat-conduction Equation (2.1.15). When deriving it, we shall speak, as in physics, about "small quantities," and so forth, without making these concepts sufficiently exact.

Consider the heat conduction in a bar of constant cross section of area S. Let the cross section be circular, to be concrete (although this assumption is not at all essential). Let the bar be insulated on its surface so that the heat cannot escape outside, and let S be small enough so that the heat conduction can be considered as one dimensional. Let ρ, c, or λ be specific mass, specific heat, or conductivity of the material, respectively. We have to derive a differential equation that the temperature $u(x, t)$ has to satisfy.

Let the x-axis be situated in the axis of the bar (see Figure 2.1.1). Let x_0 and $x_0 + \Delta x$ be interior points of the bar, x_0 be arbitrary, and $\Delta x > 0$ small. Denote by S_1 and S_2 two (circular) cross sections (of the same area S), going through the points x_0, or $x_0 + \Delta x$, respectively, and by K the cylinder between them. For its volume V, and its mass m we have, obviously

$$V = S \, \Delta x$$

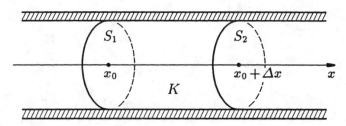

FIGURE 2.1.1

and

$$m = V\rho = S\rho \, \Delta x \tag{2.1.18}$$

respectively.

Let $t = t_0$, t_0 be arbitrary, $\Delta t > 0$ be small, and consider the time interval $I = [t_0, t_0 + \Delta t]$.

First let Δx and Δt be so small that we can assume the temperature u to be constant in K for a moment. If, during the time interval I, the temperature u increases by Δu in K, the amount of heat, needed to this aim, is

$$\Delta Q = mc \, \Delta u = S\rho \, \Delta x \, c \, \Delta u \tag{2.1.19}$$

by (2.1.18).

On the other hand, by Newton's law, through a unit area of S_2 there runs in an amount

$$+\lambda \text{ grad } u$$

of heat per unit time. However, grad $u = \partial u/\partial x$ on S_2; if we utilize, again, the fact that K is small, the amount of heat flowing into K through S_2 during the time interval I will be

$$S\lambda \frac{\partial u}{\partial x}(x_0 + \Delta x, t_0) \, \Delta t$$

Similarly, the amount of heat flowing *off* through S_1 during the same time interval is

$$S\lambda \frac{\partial u}{\partial x}(x_0, t_0) \, \Delta t$$

Through the lateral surface of K no heat enters. Thus the total amount of heat flowing into K during the time interval I is

$$S\lambda \frac{\partial u}{\partial x}(x_0 + \Delta x, t_0) \, \Delta t - S\lambda \frac{\partial u}{\partial x}(x_0, t_0) \, \Delta t \tag{2.1.20}$$

$$= S\lambda \, \Delta t \left[\frac{\partial u}{\partial x}(x_0 + \Delta x, t_0) - \frac{\partial u}{\partial x}(x_0, t_0) \right]$$

However, in accordance to the heat balance, this amount of heat should be equal to the amount ΔQ of heat needed for increasing the temperature of the cylinder K by Δu. Thus (see (2.1.19)) we have

$$S\rho\,\Delta x\,c\,\Delta u = S\lambda\,\Delta t\left[\frac{\partial u}{\partial x}(x_0 + \Delta x, t_0) - \frac{\partial u}{\partial x}(x_0, t_0)\right] \qquad (2.1.21)$$

Now, recall the mean value theorem, by which one has, for a function $f(x)$

$$f(x_0 + \Delta x) - f(x_0) = \Delta x\,f'(p)$$

where p is an interior point of the interval $(x_0, x_0 + \Delta x)$. Taking $(\partial u/\partial x)(x, t_0)$ for $f(x)$, we get

$$\frac{\partial u}{\partial x}(x_0 + \Delta x, t_0) - \frac{\partial u}{\partial x}(x_0, t_0) = \Delta x\,\frac{\partial^2 u}{\partial x^2}(p, t_0)$$

Thus the right-hand side of (2.1.21) becomes

$$S\lambda\,\Delta t\,\Delta x\,\frac{\partial^2 u}{\partial x^2}(p, t_0) \qquad (2.1.22)$$

Similarly, utilizing once more the fact that K and Δt are small, we can write

$$\Delta u = \frac{\partial u}{\partial t}(x_0, t_0)\,\Delta t$$

In this way, (2.1.19) becomes

$$S\rho\,\Delta x\,c\,\frac{\partial u}{\partial t}(x_0, t_0)\,\Delta t \qquad (2.1.23)$$

Now, replacing the left-hand, or right-hand side of (2.1.21) by (2.1.23) or (2.1.22), respectively, we obtain

$$S\rho\,c\,\frac{\partial u}{\partial t}(x_0, t_0)\,\Delta x\,\Delta t = S\lambda\,\frac{\partial^2 u}{\partial x^2}(p, t_0)\,\Delta x\Delta t \qquad (2.1.24)$$

However, $S, \rho, c, \Delta x$, and Δt are positive, so we can divide Equation (2.1.24) by their product and get

$$\frac{\partial u}{\partial t}(x_0, t_0) = \frac{\lambda}{\rho c}\frac{\partial^2 u}{\partial x^2}(p, t_0) \qquad (2.1.25)$$

Moreover, if $\Delta x \to 0$, we get $p \to x_0$, because $p \in (x_0, x_0 + \Delta x)$. Equation (2.1.25) then becomes

$$\frac{\partial u}{\partial t}(x_0, t_0) = a^2 \frac{\partial^2 u}{\partial x^2}(x_0, t_0)$$

where a^2 stands for $\lambda/(\rho c)$. However, x_0, t_0 have been chosen arbitrarily (assuming x_0 to be an interior point of the bar only), so we have

$$\frac{\partial u}{\partial t} = a^2 \frac{\partial^2 u}{\partial x^2} \tag{2.1.26}$$

everywhere.

In this way, differential Equation (2.1.15) has been derived.

REMARK 2.1.3

Let us note that the main "physical philosophy" here was the heat balance between the heat entering the "small" volume K and the heat needed for raising its temperature. Other types of equations need other physical philosophies. This is the task of theoretical physics. (See also Problem 2.4.4.)

2.2 Boundary Value Problems (Equations with Boundary Conditions); The Dirichlet Problem for Laplace and Poisson Equations; The Maximum Principle for Harmonic Functions and its Consequences

The solution of a differential equation is not uniquely determined. This could be seen in Example 2.1.1. When solving a certain mathematical, physical, or technical problem, it is necessary to have not only the corresponding differential equation, but also further information. If we have to find the deflection of a plate, for example, it is not sufficient to have knowledge of its material, loading, and so forth, which constitute the plate equation; however, it is also necessary to know whether the plate is clamped, or simply supported, and so forth. This fact leads to the formulation of the so-called *boundary conditions*. We speak about *differential equations with boundary conditions* or about so-called *boundary value problems*.

This section is devoted to the study of the Laplace and Poisson equations. Both of them belong to the simplest and, at the same time, most frequent equations encountered in applications. In this section, we shall discuss mainly the so-called *Dirichlet boundary conditions*. In this case, roughly speaking, the solution to be found should assume prescribed values on the boundary of the domain, where the given equation is investigated. However, first we have to say which types of regions we are going to consider.

Let us recall that a region in E_2 is called bounded if it is possible to find such a circle C with a sufficiently large, but finite radius, that the mentioned region can be enclosed within this circle (i.e., lies in the interior of that circle). Examples of regions that are not bounded, include the half-plane, the interior or exterior of a parabola, and so forth. Similarly, a bounded region in E_n can be defined (it can be enclosed within an N-dimensional sphere with a finite radius).

Let Ω be a region (thus an open and connected set) in E_N. *In the following text we shall always assume that Ω is a bounded region and that its boundary Γ is lipschitzian.* The concept of a lipschitzian boundary (= a Lipschitz boundary) is complicated and its precision would represent difficulties to the reader here. The reader who would like to become acquainted with exact definition, will find it in K. Rektorys 1980, [2] Chapter 28. Here we note only that regions with Lipschitz boundaries are general enough to include practically all (bounded) regions that we meet in applications. To these regions there belong, in E_2, bounded regions with smooth or piecewise smooth boundaries, without cuspidal points. In E_3 there are involved bounded regions with smooth or piecewise smooth boundaries, without singularities corresponding, in a certain sense, to cuspidal points of plane curves (edges of regression, etc.). Examples of regions of the considered type are, for $N = 2$ (thus in the plane), a circle, an annulus, a rectangle, a triangle, and so forth; in E_3 (in the three-dimensional space) a sphere, a cube, a pyramid, and so forth.

Now, let us investigate the Poisson equation

$$\Delta u = f \tag{2.2.1}$$

(and its special case, the Laplace equation

$$\Delta u = 0) \tag{2.2.2}$$

on (bounded) regions with a Lipschitz boundary. To this equation (and to Equation 2.2.2 as to its special case, which we are not going to point out again) we prescribe various boundary conditions. The following three types of them are the most current:

1. The *Dirichlet boundary condition*

$$u = g \quad \text{on} \quad \Gamma \tag{2.2.3}$$

(Figure 2.2.1), where g is a given function on the boundary Γ of the region Ω, and where Equation (2.2.1) is considered.

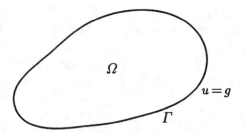

FIGURE 2.2.1

2. The *Neumann boundary condition*

$$\frac{\partial u}{\partial v} = h \quad \text{on} \quad \Gamma \tag{2.2.4}$$

Thus here the values of the derivative of the solution with respect to the outward normal v are given.

3. The *Newton boundary condition* is

$$\frac{\partial u}{\partial v} + \alpha u = k \quad \text{on} \quad \Gamma \tag{2.2.5}$$

where k is a given function and α a given constant (or function) on the boundary.

Equation (2.2.1) plus one of the given boundary conditions constitute a *problem*. Briefly we speak about the *Dirichlet*, or the *Neumann*, or the *Newton problem* for the Poisson equation.

Because the reader meets the Dirichlet problem most frequently in applications, we pay particular attention to it in this section. (At the same time, it is a model problem, where characteristic properties of Equations (2.2.1) or (2.2.2) can be especially well observed.) For the Neumann problem, which is also frequent in applications (see, e.g., K. Rektorys 1994, [1] Section 18.4; (also Problem 2.4.5). First, we make the concept of the Dirichlet problem more precise.

DEFINITION 2.2.1   (*the Dirichlet problem for the Poisson equation*).

In the region Ω let a continuous function $f(x_1, \ldots, x_N)$ be given, on Γ a continuous function g. Under the Dirichlet problem for Poisson Equation (2.2.1) we understand the problem of finding such a function $u(x_1, \ldots, x_N)$ which

1. Has, in Ω, continuous second-order partial derivatives and fulfills, in Ω, Equation 2.2.1.

2. Is continuous in the *closed* region $\overline{\Omega} = \Omega \cup \Gamma$

3. At each point of the boundary Γ satisfies the condition $u = g$

Every function with these properties is called a (*classical*) *solution* of this problem, in brief of the problem

$$\Delta u = f \quad \text{in} \quad \Omega \tag{2.2.6}$$

$$u = g \quad \text{on} \quad \Gamma \tag{2.2.7}$$

Especially, if $f(x_1, \ldots, x_N) \equiv 0$, we obtain the Dirichlet problem for the Laplace equation,

$$\Delta u = 0 \quad \text{in} \quad \Omega \tag{2.2.8}$$

$$u = g \quad \text{on} \quad \Gamma \tag{2.2.9}$$

Example 2.2.1
The function

$$u = 1 - x^2 - y^2 \tag{2.2.10}$$

is a solution of the Dirichlet problem

$$\Delta u = -4 \quad \text{in} \quad \Omega \tag{2.2.11}$$

$$u = 0 \quad \text{on} \quad \Gamma \tag{2.2.12}$$

where Ω is a circle in the plane with center at the origin and radius $R = 1$.

In fact, the function (2.2.10) has all the properties 1, 2, and 3 required in Definition 2.2.1: In Ω it has continuous second-order partial derivatives and fulfills the equation

$$\frac{\partial^2 u}{\partial x^2} + \frac{\partial^2 u}{\partial y^2} = -4$$

there, because

$$\frac{\partial^2 u}{\partial x^2} = -2$$

$$\frac{\partial^2 u}{\partial y^2} = -2$$

Moreover, it is continuous in $\overline{\Omega}$ (even in the whole xy-plane) and satisfies, on Γ, the condition in (2.2.12), since for points on Γ we have $x^2 + y^2 = 1$. ▯

Example 2.2.2
In the same way we verify that the function

$$u \equiv 1$$

is a solution of the Dirichlet problem

$$\Delta u = 0 \quad \text{in} \quad \Omega \tag{2.2.13}$$

$$u \equiv 1 \quad \text{on} \quad \Gamma \tag{2.2.14}$$

where Ω is, for example, the ellipse with semiaxes $a = 4$, $b = 2$ in the axes x, y.
▯

REMARK 2.2.1
Here we have neither theorems nor methods for constructing a solution (we become acquainted with both of them in Chapter 3). Solutions of problems (2.2.11), (2.2.12), (2.2.13), (2.2.14) were "guessed" here. A question arises, of course, whether they are unique solutions of these problems. Information about it will be obtained later (Theorem 2.2.4; see also Problem 2.4.6).

REMARK 2.2.2   *(discontinuous function f and discontinuities in boundary conditions).*
Note further that the problem formulated in Definition 2.2.1 is called *classical*, as usual ("all" should be continuous there). However, in applications we often find the case where the function f has discontinuities in the region Ω, and so does the function g on the boundary. How to formulate a solution in the first case will be shown in Chapter 3. For the second one see the conclusion of Remark 23 in Section 18.4 of the book by K. Rektorys 1994 .

In the following considerations the so-called maximum principle for harmonic functions plays an essential role.

DEFINITION 2.2.2   A function $u(x_1, \ldots, x_N)$ that has continuous second-order partial derivatives in Ω and satisfies the Laplace equation there, is called *harmonic in Ω.*

Examples of harmonic functions in every region $\Omega \subset E_2$ are the functions

$$u = x^2 - y^2$$

$$u = e^y \sin x$$

$$u = \sin x \cosh y$$

shown in Example 2.1.1.

The theorem that follows is of fundamental importance.

THEOREM 2.2.1 (the maximum principle for harmonic functions).

Let Ω be a bounded region in E_N. Let $u(x_1, \ldots, x_N)$ be a function harmonic in Ω and continuous in $\overline{\Omega}$. Denote

$$\max_{\Gamma} u = M \qquad\qquad (2.2.15)$$

$$\min_{\Gamma} u = m$$

Then everywhere in $\overline{\Omega}$ we have

$$m \overset{<}{=} u \overset{<}{=} M \qquad\qquad (2.2.16)$$

In other words, a harmonic function continuous in $\overline{\Omega}$ can assume, in Ω, neither a value higher than is its maximum on the boundary nor a value smaller than its minimum there. However, note that for a solution of the Poisson equation this theorem does not hold, for example, as is shown by the function $u = 1 - x^2 - y^2$ treated in Example 2.2.1.

The proof of Theorem 2.2.1 is rather artificial and we are not going to present it here. A "mathematical" reader finds a hint for how to prove this theorem in Problem 2.4.6.

An immediate consequence of Theorem 2.2.1 follows.

THEOREM 2.2.2
Let Ω be a bounded region in E_N, u harmonic function in Ω and continuous in $\overline{\Omega}$.
If

$$|u| \overset{\leq}{=} K \quad on \quad \Gamma \tag{2.2.17}$$

then

$$|u| \overset{\leq}{=} K \quad in \quad \overline{\Omega} \tag{2.2.18}$$

For the proof of Theorem 2.2.2 see Problem 2.4.7.

THEOREM 2.2.3
(on continuous dependence of the solution of a Dirichlet problem for the Poisson
(or Laplace) equation on boundary conditions).
Let the function u_1 be a solution of the problem

$$\Delta u = f \quad in \quad \Omega \tag{2.2.19}$$

$$u = g_1 \quad on \quad \Gamma \tag{2.2.20}$$

the function u_2 a solution of the problem

$$\Delta u = f \quad in \quad \Omega \tag{2.2.21}$$

$$u = g_2 \quad on \quad \Gamma \tag{2.2.22}$$

Let

$$|g_2 - g_1| \overset{\leq}{=} K \quad on \quad \Gamma \tag{2.2.23}$$

Then

$$|u_2 - u_1| \overset{\leq}{=} K \quad in \quad \overline{\Omega} \tag{2.2.24}$$

The PROOF is easy. According to the assumptions, u_1 satisfies

$$\Delta u_1 = f \quad in \quad \Omega \tag{2.2.25}$$

$$u_1 = g_1 \quad on \quad \Gamma \tag{2.2.26}$$

u_2 satisfies

$$\Delta u_2 = f \text{ in } \Omega \qquad (2.2.27)$$

$$u_2 = g_2 \text{ on } \Gamma \qquad (2.2.28)$$

Denote

$$u_2 - u_1 = v \qquad (2.2.29)$$

and subtract Equation (2.2.25) from Equation (2.2.27). We obtain

$$\Delta u_2 - \Delta u_1 = 0$$

that is

$$\Delta v = 0 \text{ in } \Omega \qquad (2.2.30)$$

In fact

$$\frac{\partial^2 u_2}{\partial x_1^2} - \frac{\partial^2 u_1}{\partial x_1^2} = \frac{\partial^2 (u_2 - u_1)}{\partial x_1^2} = \frac{\partial^2 v}{\partial x_1^2}$$

and similarly for other derivatives with respect to the variables $x_2, \ldots, x_N$. Thus $\Delta u_2 - \Delta u_1 = \Delta(u_2 - u_1) = \Delta v$. Moreover, the functions u_1 and u_2, being solutions of problems (2.2.19), (2.2.20), and (2.2.21), (2.2.22), are continuous in $\overline{\Omega}$ and have continuous partial derivatives of the second order in Ω, so that the same holds for the function v. Further, v satisfies (2.2.30), and thus it is harmonic in Ω. By (2.2.23) it holds

$$|v| \overset{\leq}{=} K \text{ on } \Gamma$$

Consequently, by Theorem 2.2.2, we have

$$|v| \overset{\leq}{=} K \text{ in } \overline{\Omega}$$

which is the required conclusion for (2.2.24).

REMARK 2.2.3

Let us pay attention to the fact that the function f on the right-hand sides of (2.2.19) and (2.2.21) is the same; only the functions g_1 and g_2 on the boundary are different.

From Theorem 2.2.3 it follows (if K is "small"): If two solutions of the same Poisson equation (thus with the same f), or of the Laplace equation, differ only "a little" (by K, at most) on Γ, then they differ "a little" (at most by the same K) all over $\overline{\Omega}$. This has the following practical consequence: The values of the function g

are often obtained by measurement. Thus if it is guaranteed that the values of g have been measured with an error smaller than K, then it is also guaranteed that the error of the solution is smaller than K all over $\overline{\Omega}$.

THEOREM 2.2.4 *(on uniqueness).*
The Dirichlet problem for the Poisson (or Laplace) equation cannot have more than one solution.

PROOF   Let the problem

$$\Delta u = f \ \text{ in } \ \Omega \tag{2.2.31}$$

$$u = g \ \text{ on } \ \Gamma \tag{2.2.32}$$

have two solutions u_1 and u_2, which thus satisfy

$$\Delta u_1 = f \ \text{ in } \ \Omega \tag{2.2.33}$$

$$u_1 = g \ \text{ on } \ \Gamma \tag{2.2.34}$$

and

$$\Delta u_2 = f \ \text{ in } \ \Omega \tag{2.2.35}$$

$$u_2 = g \ \text{ on } \ \Gamma \tag{2.2.36}$$

The same argumentation as in the proof of the preceding theorem (we subtract (2.2.33) from (2.2.35), etc.) leads to the conclusion that the function

$$v = u_2 - u_1$$

is harmonic in Ω and continuous in $\overline{\Omega}$. Moreover, as a consequence of (2.2.34) and (2.2.36) $v = 0$ on Γ. By the maximum principle $v \equiv 0$ in $\overline{\Omega}$, so that we really have

$$u_2 \equiv u_1 \ \text{ in } \ \overline{\Omega}$$

Especially, solutions of problems considered in Examples 2.2.1 and 2.2.2 are unique.

For the formulation of boundary conditions for the equation of plates or for the biharmonic equation see K. Rektorys 1980, [2] Chapter 23. See also Problem 2.4.2.

2.3 The Heat-Conduction Equation

Let us consider, for $x \in (0, l)$ and $t \in (0, T)$, Equation (2.1.15) for one-dimensional heat conduction without inner heat sources; thus the equation

$$\frac{\partial u}{\partial t} = a^2 \frac{\partial^2 u}{\partial x^2} \quad \text{in} \quad \Omega = (0, l) \times (0, T) \tag{2.3.1}$$

To this equation let the following conditions be prescribed (see Remark 2.3.1 later explaining their physical meaning):

$$u(x, 0) = g(x), \quad 0 < x < l \tag{2.3.2}$$

$$u(0, t) = h_1(t), \quad 0 < t \stackrel{\leq}{=} T \tag{2.3.3}$$

$$u(l, t) = h_2(t), \quad 0 < t \stackrel{\leq}{=} T \tag{2.3.4}$$

Denote by Γ the boundary of the region Ω and by Γ' this boundary with the segment

$$t = T$$

$$0 < x < l$$

excluded. In the literature, Γ' is often called the *parabolic boundary of the region* Ω. In Figure 2.3.1 this parabolic boundary is distinguished by a bold line.

DEFINITION 2.3.1   Let the function $g(x)$ be continuous in the interval $[0, l]$, and the functions $h_1(t)$ and $h_2(t)$ in the interval $[0, T]$. Moreover, let

$$h_1(0) = g(0) \tag{2.3.5}$$

$$h_2(0) = g(l)$$

be fulfilled. Then by a *(classical) solution of the problem* (2.3.1) and (2.3.4) such a function $u(x, t)$ is understood that has continuous derivatives

$$\frac{\partial u}{\partial t}$$

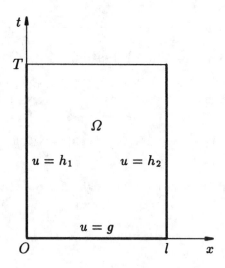

FIGURE 2.3.1

$$\frac{\partial^2 u}{\partial x^2}$$

in Ω, fulfills Equation (2.3.1) there, and is, moreover, continuous in $\overline{\Omega}$ and satisfies conditions in Equations (2.3.2) to (2.3.4) on Γ.

Example 2.3.1

The function

$$u = e^{-4t} \sin x \qquad\qquad (2.3.6)$$

is a solution of the problem

$$\frac{\partial u}{\partial t} = 4 \frac{\partial^2 u}{\partial x^2} \quad \text{in } \Omega = (0, \pi) \times (0, 1) \qquad\qquad (2.3.7)$$

$$u(x, 0) = \sin x, \quad 0 < x < \pi \qquad\qquad (2.3.8)$$

$$u(0, t) = 0, \quad 0 < t \overset{\le}{=} 1 \qquad\qquad (2.3.9)$$

$$u(\pi, t) = 0, \quad 0 < t \overset{\le}{=} 1 \qquad\qquad (2.3.10)$$

In fact, the derivatives

$$\frac{\partial u}{\partial t} = -4e^{-4t} \sin x$$

$$\frac{\partial^2 u}{\partial x^2} = -e^{-4t} \sin x$$

of this function are continuous in Ω (even in the whole x, t-plane) and evidently satisfy Equation 2.3.7. Further, the function in Equation 2.3.6 is continuous in $\overline{\Omega}$, $t = 0$ is equal to the function $\sin x$, $x = 0$, and $x = \pi$ is equal to zero. See also Problem 2.4.8. $\quad\square$

REMARK 2.3.1 *(physical interpretation of the problem (2.3.1) to (2.3.4).*

The problem (2.3.1) to (2.3.4) describes, for example, one-dimensional heat conduction in a bar of length l, insulated on its surface. This process is governed by the equation

$$\frac{\partial u}{\partial t} = a^2 \frac{\partial^2 u}{\partial x^2}$$

as has been shown in Remark 2.1.2. The process begins at the time $t = 0$, when the bar has the so-called *initial temperature* $u(x, 0)$. The condition (2.3.2) is then called the *initial condition*. The ends of the bar (at the points $x = 0$, or $x = l$) are kept at the temperatures h_1 and h_2 (conditions (2.3.3) and (2.3.4) are the so-called *boundary conditions*). The initial + boundary conditions are usually called the *mixed boundary conditions*, and the problem itself is then called the *mixed boundary value problem* for Equation (2.3.1).

In Example 2.3.1 we have $a^2 = 4$, $u(x, 0) = \sin x$, $h_1(t) \equiv 0$, and $h_2(t) \equiv 0$.

The problem (2.3.1) to (2.3.4) permits another interpretation as well: Let us consider a wall the thickness l of which (situated in the direction of the x-axis in Figure 2.3.2) is negligible as compared with its remaining dimensions (situated in the y- or z-axis). Let the initial temperature of the wall be a function of x only, let the temperature on the one face of the wall (e.g., the temperature of the room) depend on the time t only (thus be independent of y and z), and let a similar assertion be valid for the temperature on the second face of the wall (e.g., for the outside temperature). In general, for heat conduction in the wall we have the equation

$$\frac{\partial u}{\partial t} = a^2 \left(\frac{\partial^2 u}{\partial x^2} + \frac{\partial^2 u}{\partial y^2} + \frac{\partial^2 u}{\partial z^2} \right)$$

However, under these stated assumptions there is no reason for the heat to expand in the directions of the y- and z-axes. Thus, in our idealization, the temperature

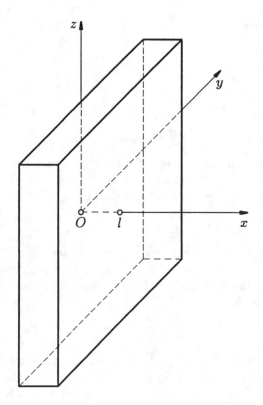

FIGURE 2.3.2

will be independent of y and z; consequently, we shall have

$$\frac{\partial^2 u}{\partial y^2} = 0$$

$$\frac{\partial^2 u}{\partial z^2} = 0$$

which implies Equation (2.3.1). Further, $g(x)$ means the initial temperature of the wall and $h_1(t)$ and $h_2(t)$ are the "inside" and "outside" temperatures.

REMARK 2.3.2

From the practical point of view the situation is a little more complicated here. In fact, on the face of the wall a thin layer of air arises, in reality, in consequence of which the boundary conditions (2.3.3) and (2.3.4) are to be replaced by the Newton conditions for heat transfer. For example, at the point $x = l$ we then have

$$\frac{\partial u}{\partial x} = -\alpha \big(u - h_2(t) \big) \tag{2.3.11}$$

where the *heat-transfer coefficient* $\alpha > 0$ depends on the material of the wall. For $\alpha \to \infty$ the condition (2.3.11) turns into the condition (2.3.4) (after dividing (2.3.11) by α and then passing to the limit for $\alpha \to \infty$). At the point $x = 0$ we shall have

$$\frac{\partial u}{\partial x} = +\alpha \big(u - h_1(t) \big) \tag{2.3.12}$$

Otherwise the situation is similar.

REMARK 2.3.3

Let us note, further, that in application conditions in (2.3.5) are very frequently not satisfied (the boundary conditions are not "compatible" with the initial temperature). Here, a comment similar to that in Remark 2.2.2 can be joint, including the quotation of Remark 23 in K. Rektorys 1994, [1] Section 18.4.

REMARK 2.3.4

Even though the heat-conduction equation is an equation of a type different from that of the Laplace one, some properties of these equations are similar. It concerns, first of all, the maximum principle and its consequences that we present here in a surveyable form only. Their proofs are almost literally the same as those presented in Section 2.2. (See also Problems 2.4.9 to 2.4.12.) Let us note that wave (2.1.16), and (2.1.17) for a vibrating string are of different character and do not have these properties. This is the reason why we do not discuss them in this chapter.

Finally, let us direct attention to Remark 2.3.6, which may be of interest.

THEOREM 2.3.1 (*maximum principle for the heat-conduction equation*). *Let $u(x, t)$ be a solution of the problem (2.3.1) to (2.3.4). Denote*

$$\max_{\Gamma'} u = M$$

$$\min_{\Gamma'} u = m$$

where Γ' is the parabolic boundary of Ω (thus Γ with the segment $t = T$, $0 < x < l$ excluded, see earlier). Then

$$m \leq u \leq M$$

all over $\overline{\Omega}$.

For the *physical meaning* of this theorem, from the point of view of physics, the maximum principle is "evident." If, in the case of an insulated bar, neither the initial temperature $g(x)$ nor the temperatures $h_1(t)$ and $h_2(t)$ of its ends exceed the value M, then the same remains valid for the temperature in the course of the whole process. (A similar conclusion holds for the minimum, of course.)

Theorem 2.3.1 has similar consequences as the maximum principle for the Laplace and Poisson equations (cf. Theorems 2.2.3 and 2.2.4).

THEOREM 2.3.2
(on continuous dependence of the solution on initial and boundary conditions).
Let the function $u_1(x, t)$ be a solution of the problem

$$\frac{\partial u}{\partial t} = a^2 \frac{\partial^2 u}{\partial x^2} \quad \text{in} \ \ \Omega = (0, l) \times (0, T)$$

$$u(x, 0) = g(x), \quad 0 < x < l$$

$$u(0, t) = h_1(t), \quad 0 < t \leqq T$$

$$u(l, t) = h_2(t), \quad 0 < t \leqq T$$

the function $u_2(x, t)$ that of the problem

$$\frac{\partial u}{\partial t} = a^2 \frac{\partial^2 u}{\partial x^2} \quad \text{in} \ \ \Omega = (0, l) \times (0, T)$$

$$u(x, 0) = \bar{g}(x), \quad 0 < x < l$$

$$u(0, t) = \bar{h}_1(t), \quad 0 < t \leqq T$$

$$u(l, t) = \bar{h}_2(t), \quad 0 < t \leqq T$$

Let

$$|\bar{g}(x) - g(x)| \leqq K \quad \text{in} \ \ [0, l]$$

$$\left|\bar{h}_1(t) - h_1(t)\right| \leqq K \quad \text{in} \ \ [0, T]$$

$$\left|\bar{h}_2(t) - h_2(t)\right| \leqq K \quad \text{in} \ \ [0, T]$$

Then we have

$$|u_2(x, t) - u_1(x, t)| \overset{\leq}{=} K$$

all over $\overline{\Omega}$.

REMARK 2.3.5
   Thus, if initial and boundary functions differ only "a little" on the boundary, then so do the corresponding solutions.

THEOREM 2.3.3 (on uniqueness).
The problem (2.3.1) to (2.3.4) cannot have more than one solution.

   Thus the function (2.3.6) is the only solution of the problem (2.3.7) to (2.3.10). See also Problem 2.4.10.

REMARK 2.3.6
   Let us note, finally, that while the maximum principle holds for the equation "without heat sources" only, Theorems 2.3.2 and 2.3.3 remain valid for the equation

$$\frac{\partial u}{\partial t} = a^2 \frac{\partial^2 u}{\partial x^2} + f(x, t)$$

as well. (See also Problems 2.4.9 and 2.4.11.)

   For various generalizations of the presented results see, for example, in K. Rektoryz 1994, [1] Section 18.6. In particular, these presented results remain valid for the N-dimensional case (cf. also Problem 2.4.12).

2.4   Problems 2.4.1 to 2.4.12

2.4.1—Show that the function

$$u = \sin x \cosh y \tag{2.4.1}$$

is a solution of the problem (see Figure 2.4.1)

$$\Delta u = 0 \quad \text{in} \quad \Omega$$

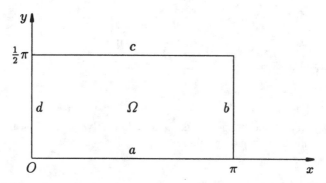

FIGURE 2.4.1

$$u = 0 \text{ on } b \text{ and } d$$

$$u = \cosh\left(\frac{1}{2}\pi\right) \sin x \text{ on } c$$

$$\frac{\partial u}{\partial v} = 0 \text{ on } a$$

where $\partial u / \partial v$ is the outward-normal derivative on a (thus $\partial u / \partial v = -\partial u / \partial y$ in our case).

2.4.2—Formulate a boundary value problem for vertical displacement of a horizontal rectangular plate, represented by a rectangular region Ω with boundary Γ, for the case when the plate is

1. Clamped

2. Simply supported

We have

$$\Delta^2 u \equiv \frac{\partial^4 u}{\partial x^4} + 2\frac{\partial^4 u}{\partial x^2 \partial y^2} + \frac{\partial^4 u}{\partial y^4} = f(x, y) \text{ in } \Omega \qquad (2.4.2)$$

(see Equation (2.1.13)),

1.

$$u = 0 \quad \text{and} \quad \frac{\partial u}{\partial v} = 0 \text{ on } \Gamma \qquad (2.4.3)$$

2.

$$u = 0 \quad \text{and} \quad \frac{\partial^2 u}{\partial v^2} = 0 \text{ on } \Gamma \tag{2.4.4}$$

(The right-hand side of ((2.4.2) characterizes vertical loading of the plate.)

2.4.3—Show that the function

$$u = \frac{q}{\pi^4 D \left(\frac{m^2}{a^2} + \frac{n^2}{b^2} \right)^2} \sin \frac{m\pi x}{a} \sin \frac{n\pi y}{b} \tag{2.4.5}$$

is the solution of problem (2.4.2) and (2.4.4) for the case that

$$\Omega = (0, a) \times (0, b)$$

and

$$f(x, y) = \frac{q}{D} \sin \frac{m\pi x}{a} \sin \frac{n\pi y}{b} \tag{2.4.6}$$

(See also Problem 5.2.8.)

2.4.4*— Derive Equation (2.1.17) of a vibrating string

$$\frac{\partial^2 u}{\partial t^2} = a^2 \frac{\partial^2 u}{\partial x^2} \tag{2.4.7}$$

Here $u(x, t)$ is vertical displacement of the string, ρ is its mass per unit length, and F is the magnitude of the tensile force F. When deriving (2.4.7), assume that u is sufficiently smooth, the displacements are small (u and its derivatives are small when compared with unity; they are so small that the increment of the length of an arc of the string over a subinterval Δx can be neglected); F has the direction of the tangent; F is constant; and the weight of the string is neglected.

Hint (see Figure 2.4.2): In contrast to heat-conduction Equation (2.1.15), where the "heat balance" was essential when deriving that equation, Newton's law of motion plays the main role here. Denote by T the center of gravity of the arc I of the string over an (interior) subinterval $[x, x + \Delta x]$; in T thus the mass of I is concentrated. Then, by Newton's law

$$\rho \, \Delta x \frac{\partial^2 u}{\partial t^2} = F \sin \beta - F \sin \alpha \tag{2.4.8}$$

However, α and β are small by assumption, so that

$$F \sin \beta - F \sin \alpha \approx F(\tan \beta - \tan \alpha) = F \left[\frac{\partial u}{\partial x} (x + \Delta x) - \frac{\partial u}{\partial x} (x) \right]$$

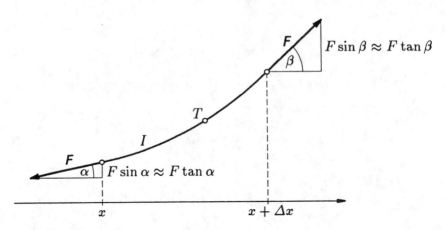

FIGURE 2.4.2

Then use the mean value theorem for the function $\partial u/\partial x$; in this way, $\partial^2 u/\partial x^2$ appears on the right-hand side of Equation (2.4.8).

2.4.5—Let the problem

$$\Delta u = f \text{ in } \Omega \tag{2.4.9}$$

$$\frac{\partial u}{\partial v} = h \text{ on } \Gamma \tag{2.4.10}$$

have a solution u_0. Show that then the function

$$u_0 + c$$

where c is an arbitrary constant, is also a solution of the same problem.

2.4.6*—Prove the maximum principle (Theorem 2.2.1) for harmonic functions.

Hint: In contradiction to Theorem 2.2.1, let such a point $P \in \Omega$ exist that $u(P) > M$. Then u attains its maximum at an *interior* point $Q(x_1^0, \ldots, x_N^0)$ of Ω, while $u(Q) > M$ again. The region Ω being bounded, such a (sufficiently small) $\alpha > 0$ exists that the function

$$v(x_1, \ldots, x_N) = u(x_1, \ldots, x_N) \tag{2.4.11}$$

$$+ \alpha \left[\left(x_1 - x_1^0 \right)^2 + \cdots + \left(x_N - x_N^0 \right)^2 \right]$$

still attains its maximum in Ω—denote that point by R. Now, R is an interior point of Ω and v attains its maximum there; thus we have necessarily

$$\frac{\partial^2 v}{\partial x_1^2}(R) \leq 0, \ldots, \frac{\partial^2 v}{\partial x_N^2}(R) \leq 0 \qquad (2.4.12)$$

However, the function u is harmonic by assumption, so that by Equation (2.4.11)

$$\frac{\partial^2 v}{\partial x_1^2}(R) + \cdots + \frac{\partial^2 v}{\partial x_N^2}(R) = \frac{\partial^2 u}{\partial x_1^2}(R) + \cdots + \frac{\partial^2 u}{\partial x_N^2}(R) + \alpha(2 + \cdots + 2)$$

$$= 0 + 2N\alpha > 0$$

in contradiction to (2.4.12). For the minimum the proof is similar.

2.4.7*— Prove Theorem 2.2.2.

Hint: From (2.2.17) we get

$$\max_{\Gamma} u = M \overset{\leq}{=} K$$

$$\min_{\Gamma} u = m \overset{\geq}{=} -K \qquad \text{on } \Gamma$$

The conclusion follows immediately from (2.2.15) and (2.2.16).

2.4.8—Show that

1. The function
$$u = \sin \frac{x}{a}$$

is a (so-called stationary) solution of the equation

$$\frac{\partial u}{\partial t} = a^2 \frac{\partial^2 u}{\partial x^2} + \sin \frac{x}{a}$$

2. The function
$$u = e^{-b^2 t} \sin \frac{x}{a}$$

is a solution of the equation

$$\frac{\partial u}{\partial t} = a^2 \frac{\partial^2 u}{\partial x^2} + (1 - b^2)e^{-b^2 t} \sin \frac{x}{a}$$

2.4.9—Theorem 2.3.2 holds true not only for the equation

$$\frac{\partial u}{\partial t} = a^2 \frac{\partial^2 u}{\partial x^2}$$

but also for the equation

$$\frac{\partial u}{\partial t} = a^2 \frac{\partial^2 u}{\partial x^2} + f(x, t)$$

Prove it!

Hint: Follow the proof of Theorem 2.2.3.

2.4.10*—Prove Theorem 2.3.3 directly, without using Theorems 2.3.1 and 2.3.2.

Hint: Let the problem (2.3.1) to (2.3.4) have two solutions $u_1(x, t)$ and $u_2(x, t)$. Denote $u = u_1 - u_2$. Then u fulfills

$$\frac{\partial u}{\partial t} = a^2 \frac{\partial^2 u}{\partial x^2} \quad \text{in} \quad \Omega$$

$$u = 0 \quad \text{on} \quad \overset{'}{\Gamma}$$

Construct the function

$$v(x, t) = e^{-\alpha t} u(x, t) \tag{2.4.13}$$

$$\alpha > 0$$

This function satisfies

$$\frac{\partial v}{\partial t} - a^2 \frac{\partial^2 v}{\partial x^2} = -\alpha v \quad \text{in} \quad \Omega \tag{2.4.14}$$

$$v = 0 \quad \text{on} \quad \overset{'}{\Gamma}$$

Now, v cannot attain a positive maximum in $\overline{\Omega} \setminus \Gamma$ (where $\overline{\Omega} \setminus \Gamma$ is $\overline{\Omega}$ with Γ excluded). In fact, at such a point we should have

$$\frac{\partial^2 v}{\partial x^2} \leqq 0$$

$$\frac{\partial v}{\partial t} \geqq 0$$

while by Equation (2.4.14)

$$\frac{\partial v}{\partial t} - a^2 \frac{\partial^2 v}{\partial x^2} = -\alpha v < 0$$

In the same way one proves that v cannot attain a negative minimum in $\overline{\Omega} \setminus \Gamma$. Thus $v \equiv 0$, and so forth.

2.4.11—Show that Theorem 2.3.3 on uniqueness also holds true for a nonhomogeneous equation

$$\frac{\partial u}{\partial t} = a^2 \frac{\partial^2 u}{\partial x^2} + f(x, t)$$

2.4.12*—The way of proof of uniqueness shown in Problem 2.4.10 can be well applied to more-dimensional cases. Formulate the problem and present the proof.

Hint: Use substitution (2.4.13) again.

Chapter 3

Variational Methods of Solution of Elliptical Boundary Value Problems; Generalized Solutions and Their Approximations; Weak Solutions

3.1 The Equation $Au = f$

In this chapter, we shall discuss equations of the form

$$Au = f \text{ in } \Omega \tag{3.1.1}$$

where A is a linear differential operator[1] (ordinary or partial), with variable coefficients in general (will be made more precise later), and $f \in L_2(\Omega)$ (see later) is a given function. The theory we are going to develop will be shown to be sufficiently general for consumers of mathematics. At the same time it is familiar to them, because it is based on the so-called theorem on minimum of functional of

---

[1] For a reader who is not familiar with the concept of an operator: In practice, operator A (or, better said, its form) is given by the left-hand side of the given differential equation. If, for example, in a two-dimensional region the equation

$$\Delta u + 2u = x^2 + y^2$$

is considered, then A in (3.1.1) is given by the relation (we say briefly "is given by")

$$Au = \Delta u + 2u$$

It means that to every fixed function $u(x, y)$ from a certain class of functions (called the domain of definition of operator A, see the next section) there corresponds the function $\Delta u + 2u$.

energy that is well known in different modifications in many physical and engineering fields. This theorem gives us a basis not only for deriving general existence theorems but also for finding effective numerical methods of solving the considered problems.

Equation 3.1.1 always will be investigated on a bounded region Ω in E_N with a lipschitzian boundary Γ. (The concept of regions with lipschitzian boundaries has been discussed in Section 2.2.) The points of the region Ω will be denoted briefly by x (instead of $(x_1, \ldots, x_N)$), in place of

$$\int \ldots \int_{\Omega} u\,(x_1, \ldots, x_N)\,dx_1 \ldots dx_N$$

we shall write

$$\int_{\Omega} u(x)dx$$

often only

$$\int_{\Omega} u\,dx$$

At the same time, the case $N = 1$ is not excluded (to be able to include ordinary differential equations as well). Then Ω is a bounded open interval and

$$\int_{\Omega} u(x)dx = \int_a^b u(x)dx$$

We denote by the symbol $C^{(j)}(\Omega)$ or $C^{(j)}(\overline{\Omega})$ the set of all functions continuous inclusive of their (partial) derivatives up to the order j in Ω or in $\overline{\Omega} = \Omega \cup \Gamma$, respectively. By the zero derivative of the function u we understand this function itself. Instead of $C^{(0)}(\Omega)$ or $C^{(0)}(\overline{\Omega})$, we write briefly $C(\Omega)$ or $C(\overline{\Omega})$, respectively. In the case $N = 1$, we use often the brief notation $C^{(j)}(a, b)$ or $C^{(j)}[a, b]$ instead of $C^{(j)}((a, b))$ or $C^{(j)}([a, b])$.

By the symbol $L_2(\Omega)$, we denote the set of functions square integrable in Ω in the Lebesgue sense (thus the integrals $\int_{\Omega} u(x)\,dx$ and $\int_{\Omega} u^2(x)\,dx$ exist and are finite), while on this set there are defined the scalar product (u, v), the norm $\|u\|$ and the distance $\rho(u, v)$ by the relations

$$(u, v) = \int_{\Omega} uv\,dx \tag{3.1.2}$$

$$\|u\| = \sqrt{(u, u)} = \sqrt{\int_{\Omega} u^2\,dx} \tag{3.1.3}$$

$$\rho(u, v) = \|v - u\| = \sqrt{\int_\Omega (v - u)^2 \, dx} \qquad (3.1.4)$$

(We also say that *metrics* has been introduced on this set.) Two functions u, v, for which

$$\rho(u, v) = 0$$

or, which is the same,

$$\sqrt{\int_\Omega (v - u)^2 \, dx} = 0$$

holds, are called *equivalent* in this space. They are considered as equal. We write

$$u = v \ \text{in} \ L_2(\Omega)$$

Equivalent functions are different at most on a set of measure zero in Ω; we say also that they are equal *almost everywhere*. If, moreover, two equivalent functions are continuous in $\overline{\Omega}$, then they are equal everywhere in $\overline{\Omega}$.

For $N = 1$ (i.e., for the space $L_2(a,b)$) see in more detail in Section 1.3.

In the sense of equality in $L_2(\Omega)$, we generalize the concept of a solution of the equation $Au = f$. We do not require that this equation be satisfied everywhere in Ω, but only in the sense of equality in $L_2(\Omega)$ (i.e., almost everywhere in Ω).

As we have already said, we assume that the equation (or, which is the same here, the differential operator A) is linear. Moreover, we shall assume that it is of an even order $2k$. In fact, this is the case most often encountered in applications. Examples of linear equations of an even order are the Laplace and Poisson equations (which are of the second order, thus $k = 1$), equations of plates and walls ($k = 2$), the ordinary differential equation

$$u'' - \left(1 + x^4\right) u = \sin x \qquad (3.1.5)$$

($k = 1$), and so forth. An example of a nonlinear equation is

$$u'' - \left(1 + x^4\right) u^2 = \sin x \ \ ^2 \qquad (3.1.6)$$

---

[2]Recall that two characteristic properties for an operator A to be linear (on its domain of definition, see Definition 3.2.1 later) are

$$A(u_1 + u_2) = Au_1 + Au_2 \qquad (a)$$

To the equation $Au = f$ we join k linear boundary conditions, which together make a "problem." For example, in the case of the Dirichlet problem for the Laplace or the Poisson equation ($k = 1$) we prescribed, in Section 2.2, the condition

$$u = g \text{ on } \Gamma \tag{3.1.7}$$

where g was a given function. For the equation of a plate ($k = 2$) we prescribe two boundary conditions, characterizing its support. If, for example, the plate is clamped, we have the conditions

$$u = 0$$

$$\frac{\partial u}{\partial \nu} = 0 \text{ on } \Gamma$$

where $\partial u / \partial \nu$ is the outward normal derivative. When the differential equation is an ordinary one, of order $2k$, we prescribe k boundary conditions at each of the endpoints a and b of the given interval. For example, for (3.1.5),

$$u'' - \left(1 + x^4\right) u = \sin x \tag{3.1.8}$$

considered on the interval $[0, 1]$, we prescribe

$$u(0) = 4, \quad u'(1) = 0 \tag{3.1.9}$$

————————————

$$A(cu) = c\, Au \qquad (b)$$

($c = $ const). Operator A_1 in (3.1.5), thus given by $A_1 u = u'' - (1 + x^4)u$, is linear, since

$$A_1 (u_1 + u_2) = (u_1 + u_2)'' - \left(1 + x^4\right)(u_1 + u_2)$$

$$= u_1'' - \left(1 + x^4\right) u_1 + u_2'' - \left(1 + x^4\right) u_2 = A_1 u_1 + A_1 u_2$$

$$A_1(cu) = (cu)'' - \left(1 + x^4\right) cu = c\left[u'' - \left(1 + x^4\right)u\right] = c\, A_1 u$$

Operator A_2 in (3.1.6), given by $A_2 u = u'' - (1 + x^4)u^2$, is *not* linear, because, for example, it does not possess the property (b). We have, in general

$$A_2(cu) = (cu)'' - \left(1 + x^4\right)(cu)^2 = cu'' - \left(1 + x^4\right) c^2 u^2 \neq c\left[u'' - \left(1 + x^4\right)u^2\right] = c\, A_2 u$$

In what follows in this chapter we shall assume that the boundary conditions are not only linear, but also homogeneous, that is, that they are satisfied by the function u identically equal to zero. (As we shall see, the advantage of this assumption lies in the fact that the set of functions, on which operator A is considered, will be a linear one.) If we prescribe, for example, boundary conditions

$$u(0) = 0, \quad u'(1) = 0 \tag{3.1.10}$$

for (3.1.8), then these conditions are not only linear, but also homogeneous. (Because the function $u \equiv 0$, having its derivative u' also identically equal to zero, fulfills the first as well as the second of the conditions (3.1.10).)

The conditions (3.1.9) (to be more precise, the first of them) are not homogeneous. However, the problem (3.1.8) and (3.1.9) can be easily transformed into that with homogeneous boundary conditions: Put

$$u = v + 4 \tag{3.1.11}$$

where v is a new unknown function. We have obviously

$$u' = v' \tag{3.1.12}$$

$$u'' = v'' \tag{3.1.13}$$

If we put now $v + 4$ for u and v'' for u'' into (3.1.8), we get the (obviously linear) differential equation for the function v

$$v'' - \left(1 + x^4\right)(v + 4) = \sin x$$

that is

$$v'' - \left(1 + x^4\right)v = \sin x + 4\left(1 + x^4\right) \tag{3.1.14}$$

From (3.1.9), (3.1.11), and (3.1.12) follow boundary conditions for the function v:

$$v(0) = 0, \quad v'(1) = 0 \tag{3.1.15}$$

The problem (3.1.14) and (3.1.15) is a problem with homogeneous boundary conditions already. Having solved it, the solution of the original problem (3.1.8) and (3.1.9) will be the function $u = v + 4$.

Even though the presented example was very, very simple, in the case of ordinary differential equations the task of transforming a problem with nonhomogeneous

boundary conditions into that with homogeneous ones does not represent difficulties. How to proceed in the case of partial differential equations (if, e.g., the
condition (3.1.7) with a nonzero function g is prescribed) can be found in K.
Rektorys 1980, [2] Chapter 11. See also Problem 3.7.9.

3.2 Comparison Functions, Domain of Definition of Operator A; Symmetrical, Positive, and Positive–Definite Operators

Let us consider, as earlier, the linear differential equation of order $2k$,

$$Au = f \ \text{ in } \ \Omega \tag{3.2.1}$$

with given linear homogeneous boundary conditions. Denote

$$D_A = \left\{ u : \ u \in C^{(2k)}(\overline{\Omega}), u \text{ satisfies the given boundary conditions} \right\} \tag{3.2.2}$$

We read it: "D_A is the set of all functions from the space $C^{(2k)}(\overline{\Omega})$ (thus of functions
continuous inclusive of their derivatives up to the order $2k$ in the closed region $\overline{\Omega}$,
see the previous section) satisfying the given boundary conditions."

DEFINITION 3.2.1

The functions from D_A are called the *comparison* (or *test*) *functions*. The set
D_A is the so-called *domain of definition of operator A*.

Let us draw the reader's attention to the fact that the domain of definition of
operator A is not region Ω, but the set of comparison functions, that is, of functions
with certain properties: The comparison functions are, first, sufficiently smooth
(so that operation Au can be realized; this is why we require $u \in C^{(2k)}(\overline{\Omega})$, because
operator A is of order $2k$). Then they fulfill the given boundary conditions. They
need not fulfill the given differential equation. Now our task will be to find,
from among all comparison functions, that one that satisfies the given equation,
since such a function, being a comparison function, fulfills the given boundary
conditions, at the same time, and is thus the wanted solution. The way we find it
will be explained in the next sections.

Example 3.2.1

Let us consider the problem

$$-u'' + \left(1 + \cos^2 x\right) u = \sin x \qquad (3.2.3)$$

$$u(0) = 0 \qquad (3.2.4)$$

$$u(\pi) = 0$$

Equation (3.2.3) is linear and of the second order. The boundary conditions are linear and homogeneous. Here, operator A is given by

$$Au = -u'' + \left(1 + \cos^2 x\right) u \qquad (3.2.5)$$

and

$$D_A = \left\{ u : u \in C^{(2)}[0, \pi], \ u(0) = 0, \ u(\pi) = 0 \right\} \qquad (3.2.6)$$

Examples of functions of D_A, thus examples of comparison functions, are then the functions

$$u = \sin x$$

$$u = x(\pi - x)$$

(Figures 3.2.1, 3.2.2). The function

$$u = \pi - x$$

(Figure 3.2.3) is not a comparison function: It belongs to $C^{(2)}[0, \pi]$ and fulfills the second of conditions (3.2.4); however, it does not satisfy the first. Not even the function sketched in Figure 3.2.4 is a comparison function. It fulfills both conditions (3.2.4), but it does not belong to $C^{(2)}[0, \pi]$. (It does not belong even to $C^{(1)}[0, \pi]$, because its first derivative is not continuous at point $\pi/2$.) ▯

REMARK 3.2.1

Let us note that D_A is a linear set. In fact, if, first, $u \in D_A$, then $cu \in D_A$ holds as well for every (real) number c, because $u \in C^{(2k)}(\overline{\Omega}) \Rightarrow cu \in C^{(2k)}(\overline{\Omega})$; if u satisfies the given boundary conditions, then cu satisfies them as well, since

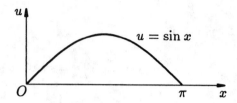

FIGURE 3.2.1

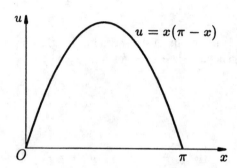

FIGURE 3.2.2

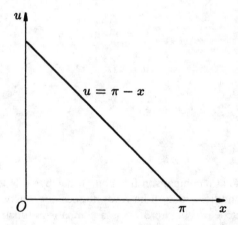

FIGURE 3.2.3

they are linear and homogeneous. A similar consideration leads to the conclusion $u \in D_A, v \in D_A \Rightarrow u + v \in D_A$.

Set D_A has been called the domain of definition of operator A, because on that set operator A is considered and its properties are investigated. Let us recall that,

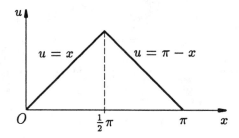

FIGURE 3.2.4

according to the assumption, A is linear on that set; thus it fulfills, on D_A, both conditions

$$A(cu) = c\, Au$$

$$A(u + v) = Au + Av$$

mentioned in the footnote on page 87.

The reader proves without difficulties that, for example, operator A given by the relation (3.2.5) is linear on its domain of definition (3.2.6) (cf. the quoted footnote).

As we said earlier, our task will be to find, from among all comparison functions, the one that satisfies the given differential equation. If we require nothing more than linearity of operator A, it is not possible to say very much about this task. Therefore, we shall restrict ourselves to a narrower class of operators, the so-called symmetrical, positive, or positive–definite ones. This class is sufficiently special to enable us to obtain relatively strong results and, at the same time, sufficiently general to involve operators most often encountered in applications. In the following section, we present corresponding definitions and show how to establish the required properties of a given operator.

DEFINITION 3.2.2

A (linear) operator A is called *Symmetrical on its domain of definition* D_A, if for every two functions $u \in D_A$ and $v \in D_A$ the relation

$$(Au, v) = (u, Av) \tag{3.2.7}$$

holds

Positive on D_A if it is symmetrical and if, moreover

$$(Au, u) \overset{\geq}{=} 0 \quad \text{holds for every } u \in D_A \tag{3.2.8}$$

while

$$(Au, u) = 0 \implies u \equiv 0 \qquad (3.2.9)$$

Positive definite on D_A if it is symmetrical and if there exists a constant $C > 0$, independent of u, so that for every comparison function $u \in D_A$ we have

$$(Au, u) \overset{\geq}{=} C^2 \|u\|^2 \qquad (3.2.10)$$

where $\|u\|$ is the norm of the function u in the space $L_2(\Omega)$ (see (3.1.3)).

REMARK 3.2.2

Every positive or positive–definite operator is symmetrical by definition. A not very complicated consideration leads to the conclusion that every positive–definite operator is positive.

Example 3.2.2

Let us consider operator A from Example 3.2.1, given on the set of comparison functions (3.2.6), that is, on the set

$$D_A = \left\{ u : u \in C^{(2)}[0, \pi], \ u(0) = 0, \ u(\pi) = 0 \right\} \qquad (3.2.11)$$

by

$$Au = -u'' + \left(1 + \cos^2 x\right) u \qquad (3.2.12)$$

We show that this operator is symmetrical and positive on D_A, even positive–definite.

Symmetry — We have to prove (Equation 3.2.7), that is

$$u, v \in D_A \implies (Au, v) = \int_0^\pi \left[-u'' + (1 + \cos^2 x)u \right] v \, dx \qquad (3.2.13)$$

$$= (u, Av) = \int_0^\pi u \left[-v'' + \left(1 + \cos^2 x\right) v \right] dx$$

First, we have obviously

$$\int_0^\pi \left(1 + \cos^2 x\right) uv \, dx = \int_0^\pi u \left(1 + \cos^2 x\right) v \, dx \qquad (3.2.14)$$

Thus it remains to be proved that

$$u, v \in D_A \implies \int_0^\pi \left(-u''\right) v \, dx = \int_0^\pi u \left(-v''\right) dx \tag{3.2.15}$$

However, if $u \in D_A$ and $v \in D_A$, then by (3.2.11)

$$u(0) = 0, \quad u(\pi) = 0, \quad v(0) = 0, \quad v(\pi) = 0 \tag{3.2.16}$$

Thus, integrating by parts, we obtain

$$-\int_0^\pi u'' v \, dx = -\left[u'v\right]_0^\pi + \int_0^\pi u'v' \, dx = \int_0^\pi u'v' \, dx \tag{3.2.17}$$

$$= \left[uv'\right]_0^\pi - \int_0^\pi uv'' \, dx = -\int_0^\pi uv'' \, dx \tag{3.2.18}$$

$$= \int_0^\pi u \left(-v''\right) dx$$

In this way, (3.2.15) is proved. From (3.2.14) and (3.2.15) there follows symmetry of operator A on D_A.

At the same time, (3.2.17) implies

$$(Au, v) = \int_0^\pi \left[-u'' + \left(1 + \cos^2 x\right) u\right] v \, dx \tag{3.2.19}$$

$$= \int_0^\pi \left[u'v' + \left(1 + \cos^2 x\right) uv\right] dx$$

Positivity — Symmetry of operator A on D_A has just been proved. Thus we have to prove (3.2.8) and (3.2.9), that is,

$$(Au, u) \overset{\geq}{=} 0 \quad \text{for every } u \in D_A \tag{3.2.20}$$

$$(Au, u) = 0 \implies u \equiv 0 \tag{3.2.21}$$

However, by (3.2.19), where we put $v = u \in D_A$, we obtain

$$(Au, u) = \int_0^\pi \left[u'^2 + \left(1 + \cos^2 x\right) u^2\right] dx \overset{\geq}{=} 0 \tag{3.2.22}$$

It remains to prove (3.2.21). Thus let $(Au, u) = 0$, that is, let

$$\int_0^\pi u'^2 \, dx + \int_0^\pi \left(1 + \cos^2 x\right) u^2 \, dx = 0 \tag{3.2.23}$$

However, each of the integrals in (3.2.23) is nonnegative; consequently (3.2.23) is fulfilled only if each of them is equal to zero. Further, $u \in C^{(2)}[0, \pi]$, so that u' is a continuous function in $[0, \pi]$. Thus, taking the first of the integrals in (3.2.23) into consideration, we obtain

$$\int_0^\pi u'^2 \, dx = 0 \implies u' \equiv 0 \implies u \equiv \text{const in } [0, \pi] \tag{3.2.24}$$

However, $u \in D_A$, so $u(0) = 0$, and therefore $u \equiv 0$. In this way (3.2.21) is proved. Positivity of operator A is established.

Positive Definiteness — Symmetry of operator A was proved earlier. We have to prove (3.2.10), that is, existence of such a number $C > 0$, independent of u, that

$$u \in D_A \implies (Au, u) \overset{\geq}{=} C^2 \|u\|^2 \tag{3.2.25}$$

holds, where

$$\|u\|^2 = \int_0^\pi u^2 \, dx$$

However, we have

$$(Au, u) = \int_0^\pi \left[u'^2 + \left(1 + \cos^2 x\right) u^2 \right] dx \overset{\geq}{=} \int_0^\pi u^2 \, dx \tag{3.2.26}$$

from which immediately follows (3.2.25) with $C = 1$.
In this way positive definiteness of operator A is also proved. □

REMARK 3.2.3

From this proved positive definiteness there follows positivity of operator A (see Remark 3.2.2). The preceding proof of positivity was thus "superfluous." However, positive definiteness could be proved so easily here because the coefficient at u in (3.2.12), that is, the function $r(x) = 1 + \cos^2 x$, was sharply positive here, $r(x) \geq \text{const} > 0$ in $[0, \pi]$. For example, in the case of the problem

$$-u'' + xu = 1, \quad u(0) = 0, \quad u(1) = 0$$

the function $r(x) = x$ is only nonnegative in $[0, 1]$, and it is not possible to apply the straightforward way shown in (3.2.26). Here, the proof of positive definiteness is not as easy (see, e.g., A. Zěnísěk 1990, [2] Chapter 8), while the proof of positivity, which is often sufficient for our considerations, remains unchanged.

REMARK 3.2.4

In Definition 3.2.2 we speak of symmetry, positivity and positive definiteness of operator A on its domain of definition, that is, on the set of the comparison functions. Let us note that, when establishing symmetry of the operator in (3.2.12), we essentially utilized the fact that u and v were comparison functions: These functions fulfill the conditions (3.2.16), so that, by integrating by parts, both expressions

$$\left[u'v\right]_0^\pi = u'(\pi)v(\pi) - u'(0)v(0)$$

and

$$\left[uv'\right]_0^\pi = u(\pi)v'(\pi) - u(0)v'(0)$$

vanish, and this leads to the required result. The mere smoothness of the functions u and v is not sufficient. It is easy to verify that the operator, given by (3.2.12) and symmetrical on the set of functions in (3.2.11) (as we have proved), is not symmetrical, for example, on the set $C^{(2)}[0, \pi]$. Functions of that set are sufficiently smooth, but they need not be equal to zero at the endpoints of the interval $[0, \pi]$. If we take, for example, the functions

$$u = x$$

$$v = x(x - \pi)$$

each of which belongs to $C^{(2)}[0, \pi]$, we obtain

$$(Au, v) = \int_0^\pi \left[-u'' + \left(1 + \cos^2 x\right) u\right] v \, dx = \int_0^\pi \left[\left(1 + \cos^2 x\right) x\right] x(x-\pi) \, dx$$

because $u'' \equiv 0$, while

$$(u, Av) = \int_0^\pi u \left[-v'' + \left(1 + \cos^2 x\right) v\right] dx$$

$$= \int_0^\pi x \cdot (-2) \, dx + \int_0^\pi \left(1 + \cos^2 x\right) x \cdot x(x - \pi) \, dx$$

$$= -\pi^2 + \int_0^\pi \left[\left(1 + \cos^2 x\right) x\right] x(x - \pi) \, dx \neq (Au, v)$$

Thus the fact that operator A is investigated on its domain of definition D_A, the elements (= functions) that fulfill, by definition, the given boundary conditions, is essential.

In the case of *partial* differential equations, the proof of symmetry, and so forth is more complicated.

Example 3.2.3
On a bounded region Ω with a Lipschitz boundary consider the problem

$$-\Delta u = f \text{ in } \Omega \tag{3.2.27}$$

$$u = 0 \text{ on } \Gamma \tag{3.2.28}$$

In our case we have

$$D_A = \left\{ u : u \in C^{(2)}(\overline{\Omega}), \ u = 0 \text{ on } \Gamma \right\} \tag{3.2.29}$$

We show that operator A, given by

$$Au = -\Delta u = -\left(\frac{\partial^2 u}{\partial x_1^2} + \cdots + \frac{\partial^2 u}{\partial x_N^2} \right) \tag{3.2.30}$$

is symmetrical and positive on D_A.

Instead of the formula for integration by parts for $N = 1$, we use here the so-called *Green formula* (often also called *formula for integration by parts for functions of several variables*)

$$\int_\Omega \frac{\partial g}{\partial x_i} h \, dx = \int_\Gamma gh \, v_i \, dS - \int_\Omega g \frac{\partial h}{\partial x_i} \, dx, \quad i = 1, \ldots, N \tag{3.2.31}$$

valid for functions g and $h \in C^{(1)}(\overline{\Omega})$ (assuming Ω to be bounded in E_N, with a Lipschitz boundary). The first integral on the right-hand side of Equation (3.2.31) is a surface integral (for $N = 2$ a curvilinear integral) over the boundary Γ, v_i is the ith component of the unit outward normal, thus its ith direction cosine. Let

$$u \in D_A, \quad v \in D_A \tag{3.2.32}$$

Then, first

$$\frac{\partial u}{\partial x_i} \in C^{(1)}(\overline{\Omega}), \quad v \in C^{(2)}(\overline{\Omega})$$

Thus we can put, in (3.2.31)

$$g = \frac{\partial u}{\partial x_i}, \quad h = v$$

Further, $v \in D_A \implies v = 0$ on Γ by (3.2.29). Thus we obtain for the functions (3.2.32)

$$\int_\Omega \frac{\partial^2 u}{\partial x_i^2} v \, dx = \int_\Gamma \frac{\partial u}{\partial x_i} v \, v_i \, dS - \int_\Omega \frac{\partial u}{\partial x_i} \frac{\partial v}{\partial x_i} \, dx = - \int_\Omega \frac{\partial u}{\partial x_i} \frac{\partial v}{\partial x_i} \, dx \quad (3.2.33)$$

▯

Symmetry of Operator A (3.2.30) on D_A —

$$(Au, v) = - \int_\Omega \left(\frac{\partial^2 u}{\partial x_1^2} + \ldots + \frac{\partial^2 u}{\partial x_N^2} \right) v \, dx \qquad (3.2.34)$$

$$= \int_\Omega \left(\frac{\partial u}{\partial x_1} \frac{\partial v}{\partial x_1} + \ldots + \frac{\partial u}{\partial x_N} \frac{\partial v}{\partial x_N} \right) dx$$

by (3.2.33). Similarly

$$(u, Av) = - \int_\Omega u \left(\frac{\partial^2 v}{\partial x_1^2} + \ldots + \frac{\partial^2 v}{\partial x_N^2} \right) dx$$

$$= \int_\Omega \left(\frac{\partial v}{\partial x_1} \frac{\partial u}{\partial x_1} + \ldots + \frac{\partial v}{\partial x_N} \frac{\partial u}{\partial x_N} \right) dx$$

In this way, symmetry of operator A on D_A is proved.

Positivity — Symmetry has just been proved. In accordance with Definition 3.2.2 we have to prove

$$u \in D_A \implies (Au, u) \overset{\geq}{=} 0 \qquad (3.2.35)$$

while

$$(Au, u) = 0 \implies u \equiv 0 \text{ in } \overline{\Omega} \qquad (3.2.36)$$

However, by (3.2.34) we have, putting $v = u$

$$(Au, u) = \int_\Omega \left[\left(\frac{\partial u}{\partial x_1} \right)^2 + \ldots + \left(\frac{\partial u}{\partial x_N} \right)^2 \right] dx \qquad (3.2.37)$$

which immediately implies (3.2.35). Further $\partial u / \partial x_i \in C^{(1)}(\overline{\Omega})$, so that (using (3.2.37))

$$(Au, u) = 0 \implies \frac{\partial u}{\partial x_i} \equiv 0 \quad \text{for } i = 1, \ldots, N$$

from which $u \equiv \text{const in } \overline{\Omega}$. However, $u \in D_A$, thus $u = 0$ on Γ, and, consequently $u \equiv 0$ in $\overline{\Omega}$. Both conditions (Equations 3.2.35 and 3.2.36) are fulfilled. The operator (Equation 3.2.30) is positive on D_A.

The proof of positive definiteness is based on the so-called Friedrichs inequality and is more complicated (see, e.g., K. Rektorys 1980, [2] Chapter 22).

REMARK 3.2.5   *(table of positive definite problems)*

For the reader's convenience we present a small table of positive definite problems. (It is usual to speak about a *symmetrical, positive,* or *positive–definite problem* instead of a problem with a symmetrical, positive, or positive–definite operator, respectively.) Extensive tables can be found in K. Rektorys 1994, [1] Chapter 24 and in K. Rektorys 1980, [2].

In Table 3.1, conditions for positive definiteness are given in the form of inequalities. Assumptions on smoothness can be easily completed by the reader; $f \in L_2(\Omega)$ is assumed.

On the basis of Table 3.1 it is possible to decide immediately on positive definiteness of a lot of problems. Consider, for example, the problem (3.2.3) and (3.2.4),

$$-u'' + \left(1 + \cos^2 x \right) u = \sin x \qquad (3.2.38)$$

$$u(0) = 0 \qquad (3.2.39)$$

$$u(\pi) = 0$$

In the notation of Problem 1 in Table 3.1 we have $a = 0$, $b = \pi$, $p(x) \equiv 1 > 0$, $r(x) = 1 + \cos^2 x \geq 0$, which implies (by this table) positive definiteness of the problem (3.2.38) and (3.2.39). Also the problem (3.2.27) and (3.2.28) is contained in Table 3.1 (Problem 5).

Table 3.1   Positive Definite Problems

Problem	Equation	Boundary Conditions
1	$-(pu')' + ru = f,$	$u(a) = 0, \ u(b) = 0$
	$p(x) \overset{\geq}{=} p_0 > 0, \ r(x) \overset{\geq}{=} 0$	
2	The same	$u(a) = 0, \ u'(b) = 0$
3	$(pu'')'' - (gu')' + ru = f,$	$u(a) = 0, \ u(b) = 0,$
	$p(x) \overset{\geq}{=} p_0 > 0, \ g(x) \overset{\geq}{=} 0,$	$u'(a) = 0, \ u'(b) = 0$
	$r(x) \overset{\geq}{=} 0$	
4	The same	$u(a) = 0, \ u(b) = 0,$
		$u''(a) = 0, \ u''(b) = 0$
5	$-\Delta u = f$	$u = 0$ on Γ
6	$\Delta^2 u = f$	$u = 0, \ \dfrac{\partial u}{\partial \nu} = 0$ on Γ

Let us note, further, that if we write differential Equation (3.2.38) in the form

$$u'' - \left(1 + \cos^2 x\right) u = -\sin x \tag{3.2.40}$$

then the corresponding operator A, given by

$$Au = u'' - \left(1 + \cos^2 x\right) u \tag{3.2.41}$$

will not be positive (and the less positive definite) on the set

$$D_A = \left\{ u : u \in C^{(2)}[0, \pi], \ u(0) = 0, \ u(\pi) = 0 \right\} \tag{3.2.42}$$

This problem can be easily converted into the problem (3.2.38) and (3.2.39), of course, multiplying Equation (3.2.40) by the number -1. Thus it follows that "good properties" of a problem can often be attained when writing it in a suitable form.

REMARK 3.2.6   *(to the concept of positivity)*
Positivity or positive definiteness of a problem is a certain "criterion" of its reasonability. In the introduction, we encountered the problems

$$-u'' = -2 \tag{3.2.43}$$

$$u'(0) = 0, \quad u'(1) = 0 \tag{3.2.44}$$

and

$$-u'' = 0 \tag{3.2.45}$$

$$u'(0) = 0, \quad u'(1) = 0 \tag{3.2.46}$$

(here we write them in a form suitable for further investigation, thus with the original equations multiplied by the number -1). In the introduction we showed that the first of these problems had no solution, while the second one had an infinite number of solutions (of the form $u = $ const.). These "unreasonable" results are connected with the fact that the corresponding operator A, given by

$$Au = -u''$$

on the set

$$D_A = \left\{ u : u \in C^{(2)}[0, 1], \ u'(0) = 0, \ u'(1) = 0 \right\}$$

is not positive on this set: Symmetry can be easily shown, because for arbitrary functions $u \in D_A$ and $v \in D_A$ we have

$$u'(0) = 0, \quad u'(1) = 0, \quad v'(0) = 0, \quad v'(1) = 0$$

and, consequently

$$(Au, v) = -\int_0^1 u'' v \, dx = -\left[u' v \right]_0^1 + \int_0^1 u' v' \, dx = \int_0^1 u' v' \, dx \tag{3.2.47}$$

$$= \left[u v' \right]_0^1 - \int_0^1 u v'' \, dx = -\int_0^1 u v'' \, dx = (u, Av) \tag{3.2.48}$$

However, positivity is not at hand. In fact, let

$$(Au, u) = 0$$

that is, let

$$\int_0^1 u'^2 \, dx = 0 \tag{3.2.49}$$

(by (3.2.47)). Then (3.2.49) yields

$$u' \equiv 0 \quad \text{and} \quad u \equiv \text{const} \tag{3.2.50}$$

as in Example 3.2.2, but nothing more, because, in contrast to that example, the conditions $u'(0) = 0$ and $u'(1) = 0$ do not imply that the constant (3.2.50) must be equal to zero (in fact, it may be arbitrary). Thus, the condition in (3.2.9) is not fulfilled, and, consequently, operator A is not positive on D_A.

In the following we shall see that positive operators have "reasonable" properties.

THEOREM 3.2.1 *(on uniqueness).*

The equation

$$Au = f$$

where A is a positive operator on its domain of definition D_A, cannot have more than one solution.

The PROOF is easy: Let two such solutions exist, that is, two functions u_1 and $u_2 \in D_A$ satisfying, in $L_2(\Omega)$

$$Au_1 = f \tag{3.2.51}$$

$$Au_2 = f \tag{3.2.52}$$

Denote $u_2 - u_1 = v$ and subtract (3.2.51) from (3.2.52). We obtain

$$Au_2 - Au_1 = 0$$

and, by linearity of operator A

$$A(u_2 - u_1) = 0$$

that is

$$Av = 0 \tag{3.2.53}$$

By multiplying (3.2.53) scalarly by the function v, we get

$$(Av, v) = 0 \;^3 \tag{3.2.54}$$

However, operator A is positive, by assumption, and this implies

$$v = 0$$

that is

$$u_2 = u_1$$

that which was to be proved.

3.3   Theorem on Minimum of Functional of Energy

Let us consider our equation

$$Au = f \tag{3.3.1}$$

where $f \in L_2(\Omega)$ is a given function and A is an operator positive on D_A. Denote

$$Fu = (Au, u) - 2(f, u) = \int_\Omega Au \cdot u \, dx - 2 \int_\Omega fu \, dx, \quad u \in D_A \tag{3.3.2}$$

If $u \in D_A$ is a fixed function, then each of the integrals, that is, of the scalar products in (3.3.2), is a certain (real) number, depending on how the function $u \in D_A$ has been chosen. (The function f is fixed, being the given right-hand side of (3.3.1).) So by (3.3.2), to each function $u \in D_A$, a certain number is assigned. Such a relation, by which to every function u of a given set there corresponds a certain number, is called a *functional* (given on that set). Consequently, (3.3.2) is a functional given on set D_A of comparison functions. This functional is often called *functional of energy* (often also *quadratic functional*), for reasons to be seen later (see, especially, Example 3.3.1).

---

[3] A note for the reader wanting to have a better insight into the proof: By assumption, (3.2.51) and (3.2.52) are to be fulfilled in $L_2(\Omega)$, that is, almost everywhere in Ω. Consequently, (3.2.53) is fulfilled in the same sense. Now, we multiply (3.2.53) scalarly by the function v, that is, we multiply the left-hand side as well as the right-hand side of this equation by the function v and integrate over Ω. On the left-hand side we thus obtain the scalar product (Av, v). On the right-hand side we multiply, by v, a function that is equal to zero almost everywhere. Thus we get a function almost equal to zero, and the integral of such a function is equal to zero.

Now, we are going to formulate the fundamental theorem of this section. Its main significance lies in the fact that the problem of finding, among all the comparison functions, one that is the solution of the equation $Au = f$ can be replaced by the problem of finding such a function that minimizes, on D_A, the functional of energy. For solving the latter problem then efficient methods can be applied.

THEOREM 3.3.1 *(on minimum of functional of energy)*

Let us consider the equation

$$Au = f \tag{3.3.3}$$

with a positive operator A and let

$$Fu = (Au, u) - 2(f, u) \tag{3.3.4}$$

be the corresponding functional of energy. We have

1. If $u_0 \in D_A$ is a solution of Equation (3.3.3), then for this function the functional (3.3.4) attains its minimum on D_A. Moreover, this minimum is a sharp one: If $u \in D_A$, $u \neq u_0$, then $Fu > Fu_0$.

2. If, for a function $u_0 \in D_A$, the functional (3.3.4) attains its minimum on D_A, then u_0 is the solution (the only one by Theorem 3.2.1) of Equation (3.3.3).

Because Theorem 3.3.1 is of fundamental importance, we present the PROOF.

First, let $u_0 \in D_A$ be a solution of Equation (3.3.3), that is, let

$$Au_0 = f \text{ in } \Omega \tag{3.3.5}$$

hold (with the possible exception of a set of zero measure). Then

$$Fu = (Au, u) - 2(f, u) = (Au, u) - 2(Au_0, u) \tag{3.3.6}$$

by (3.3.5). If we utilize symmetry of the scalar product first, and then symmetry of operator A (recall that a positive operator is symmetrical by definition), we have

$$(Au_0, u) = (u, Au_0) = (Au, u_0)$$

The functional (3.3.6) then becomes

$$Fu = (Au, u) - 2(Au_0, u) \tag{3.3.7}$$

$$= (Au, u) - (Au_0, u) - (Au_0, u)$$

$$= (Au, u) - (Au_0, u) - (Au, u_0)$$

$$= (Au, u) - (Au_0, u) - (Au, u_0) + (Au_0, u_0) - (Au_0, u_0)$$

$$= (A(u - u_0), u - u_0) - (Au_0, u_0)$$

Thus

$$Fu = (A(u - u_0), u - u_0) - (Au_0, u_0) \qquad (3.3.8)$$

or, with the notation

$$u - u_0 = z, \quad Fu = (Az, z) - (Au_0, u_0) \qquad (3.3.9)$$

However, A is positive on D_A; consequently

$$(Az, z) = 0 \quad \text{for } z = 0$$

$$(Az, z) > 0 \quad \text{for } z \neq 0$$

Thus if $u = u_0$, then

$$Fu = Fu_0 = -(Au_0, u_0)$$

and if $u \neq u_0$, then

$$Fu > -(Au_0, u_0)$$

Thus the functional F attains its minimum exactly for the function u_0, which is the solution of Equation (3.3.3).

Second, let u_0 be a function that minimizes the functional (3.3.4) on D_A, that is, let

$$\min_{u \in D_A} Fu = Fu_0 \qquad (3.3.10)$$

Let us choose an arbitrary, but fixed function $v \in D_A$ and let t be an arbitrary (real) number. Because D_A is a linear set, $u = u_0 + tv$ belongs to D_A again. Moreover, by (3.3.10)

$$F(u_0 + tv) \geq Fu_0 \qquad (3.3.11)$$

If we use, as earlier, linearity and symmetry of operator A and symmetry of the scalar product, we obtain

$$F(u_0 + tv) = (A(u_0 + tv), u_0 + tv) - 2(f, u_0 + tv)$$

$$= (Au_0 + tAv, u_0 + tv) - 2(f, u_0) - 2t(f, v)$$

$$= (Au_0, u_0) + t(Av, u_0) + t(Au_0, v)$$

$$+ t^2(Av, v) - 2t(f, v) - 2(f, u_0)$$

$$= (Au_0, u_0) + 2t(Au_0, v) + t^2(Av, v) - 2t(f, v) - 2(f, u_0)$$

Thus we have

$$F(u_0 + tv) = (Au_0, u_0) + 2t(Au_0, v) \tag{3.3.12}$$

$$+ t^2(Av, v) - 2t(f, v) - 2(f, u_0)$$

Now, u_0 and f are fixed functions, and (3.3.12) shows that for every fixed $v \in D_A$ the functional $F(u_0 + tv)$ is a quadratic function of the variable t. According to (3.3.10), this function has to have a local minimum for $t = 0$. Consequently, its first derivative with respect to t should be equal to zero for $t = 0$, that is

$$\left. \frac{d}{dt} F(u_0 + tv) \right|_{t=0} = 0 \tag{3.3.13}$$

Thus differentiating the right-hand side of (3.3.12) with respect to t

$$2(Au_0, v) + 2t(Av, v) - 2(f, v)$$

and putting the result equal to zero for $t = 0$ by (3.3.13), we obtain

$$2(Au_0, v) - 2(f, v) = 0$$

This condition can obviously be written in the form

$$(Au_0 - f, v) = 0 \tag{3.3.14}$$

However, the function $v \in D_A$ has been chosen arbitrarily, so that (3.3.14) holds for every $v \in D_A$. From this it follows that $Au_0 - f$ is equal to zero in $L_2(\Omega)^4$,

---

[4] Here we have used a well-known result from the theory of space $L_2(\Omega)$: If an element $z \in L_2(\Omega)$ is orthogonal to every element of a set M that is dense in $L_2(\Omega)$, then z is the zero element in $L_2(\Omega)$. Proof of density of D_A in $L_2(\Omega)$ for problems considered in this book can be found in K. Rektorys 1980, [2] Chapter 8.

that is

$$Au_0 - f = 0 \text{ in } L_2(\Omega)$$

Thus, u_0 is actually a solution of Equation (3.3.3).

This completes the proof of Theorem 3.3.1.

Example 3.3.1
(deflection of a bar on an elastic foundation).
   Let us consider the equation

$$\left(EIu''\right)'' + ru = q \tag{3.3.15}$$

with boundary conditions

$$u(0) = 0, \quad u(l) = 0, \quad u'(0) = 0, \quad u'(l) = 0 \tag{3.3.16}$$

Let us assume that the functions $E(x)$ and $I(x)$, and their derivatives of the first and second orders, as well as the function $r(x)$ are continuous in the interval $[0, l]$, while

$$E(x) \stackrel{\geq}{=} \text{const} > 0, \quad I(x) \stackrel{\geq}{=} \text{const} > 0 \tag{3.3.17}$$

$$r(x) \stackrel{\geq}{=} 0, \quad q \in L_2(0, l)$$

   This problem can be interpreted as the problem of finding the deflection of a horizontal bar of length l, nonhomogeneous in general and with a variable cross section (modulus of elasticity E as well as moment of inertia I of the cross section with respect to the axis of the bar depend on x, in general), on an elastic foundation (the function r characterizes stiffness of the foundation and may depend on x as well), clamped at its ends (conditions (3.3.16)), with a vertical load $q(x)$.

   According to Table 3.1, Problem 3, our problem is positive definite, under the given assumptions, and thus positive as well. (Symmetry and positivity of operator A, given by

$$Au = \left(EIu''\right)'' + ru \tag{3.3.18}$$

on the set of comparison functions

$$D_A = \tag{3.3.19}$$

$$\{u : \ u \in C^{(4)}[0, l], \ u(0) = 0, \ u(l) = 0, \ u'(0) = 0, \ u'(l) = 0\}$$

can be easily derived by the reader: Repeated integration by parts and conditions (3.3.16) give

$$u, v \in D_A \implies (Au, v) = \int_0^l EIu''v'' \, dx + \int_0^l ruv \, dx = (u, Av)$$

consequently,

$$(Au, u) \geqq 0$$

$$(Au, u) = 0 \implies \int_0^l EIu''^2 \, dx = 0 \implies u'' \equiv 0 \implies u = c_1 x + c_2$$

and $c_1 = 0$ and $c_2 = 0$ follow by (3.3.16).)

The functional F is of the form here

$$Fu = (Au, u) - 2(q, u) \tag{3.3.20}$$

$$= \int_0^l EIu''^2 \, dx + \int_0^l ru^2 \, dx - 2 \int_0^l qu \, dx$$

$$= 2 \left\{ \frac{1}{2} \int_0^l EIu''^2 \, dx + \frac{1}{2} \int_0^l ru^2 \, dx - \int_0^l qu \, dx \right\}$$

and gives—when speaking physically—double total energy of the considered bar. (As well known, the integral

$$\frac{1}{2} \int_0^l EIu'' \, dx$$

or

$$\frac{1}{2} \int_0^l ru^2 \, dx$$

or

$$- \int_0^l qu \, dx$$

gives—for a fixed u—elastic energy of the bar, or resistance energy of the foundation, or potential energy of the vertical loading, respectively.) Physically, or technically speaking, the theorem on minimum of functional of energy has the following meaning here: From all "virtual" deflections, that is, from all (sufficiently smooth) deflections respecting geometrical boundary conditions (3.3.16), the wanted deflection (thus satisfying the differential (3.3.15)) is just that for which the functional (3.3.20) attains its minimum. $\square$

The presented example shows a close connection between the theorem on minimum of functional of energy and variational principles of mechanics. However, Theorem 3.3.1 is a general mathematical theorem, which may have a much broader application than in mechanics only.

REMARK 3.3.1

As we have mentioned, the main purpose of Theorem 3.3.1 lies in the fact that on its basis it is possible to convert the problem of finding, on set D_A of comparison functions, such a function that fulfills the given equation, in the problem of finding a function for which the functional F is minimal on that set. Some methods to solve the latter problem will be shown in Section 3.5. First we have to mention the following fact: Theorem 3.3.1 is of a conditional character. It asserts, namely: If there exists such a function $u_0 \in D_A$ that is a solution of the equation $Au = f$, then this function minimizes the functional F on D_A. Conversely: If there exists such a function $u_0 \in D_A$ for which the functional F attains its minimum on D_A, then this function is a solution of the equation $Au = f$. However, Theorem 3.3.1 is not an "existence theorem." It does not assert, namely, that a function $u_0 \in D_A$ exists that is a solution of the equation $Au = f$ on D_A, or that the functional F attains its minimum on D_A for a certain function $u_0 \in D_A$. In fact, such a function does not need to exist even in very simple cases. It is sufficient to consider Example 3.3.1 with the function q "sufficiently discontinuous." (It suffices if, e.g., q is piecewise constant in $[0, l]$.) In fact, for every function $u_0 \in D_A$, the left-hand side of Equation (3.3.15) is a continuous function in $[0, l]$, since $u_0 \in D_A$ implies $u_0 \in C^{(4)}[0, l]$, and the functions E, E', E'', I, I', I'', and r are continuous in $[0, l]$ by assumption. Now, a continuous function and a discontinuous function of the previously mentioned type cannot be equivalent in $L_2(0, l)$. Thus, no function $u_0 \in D_A$ can be a solution of Equation (3.3.15). Consequently, no function $u_0 \in D_A$ can minimize the functional F on D_A, since then this function should be a solution of the given equation by Theorem 3.3.1.

It is clear, by this example, that it is necessary to extend the concept of a solution of the given equation in a certain way, since otherwise we would not be able to include, in our theory, even very simple cases arising in applications (deflection of a bar loaded discontinuously, etc.). How to do it is relatively simple: Extend the domain of definition D_A in a proper way and extend, simultaneously, the functional F to that extended domain. If we succeed in extending D_A so that the extended functional attains its minimum for a certain function u_0, we will call this function a *generalized solution* of the given problem. Such an extension can be obtained, in a relatively simple way, if operator A is positive definite on D_A. How to do it is shown in K. Rektorys 1980, [2] Chapter 10, using tools of functional analysis. However, the present book has been written primarily for engineers and scientists, and functional analysis is a branch of mathematics that may be rather difficult for them. Thus, we choose another approach, making it possible to introduce further concepts, currently used in variational methods, especially with generalized derivatives, the Sobolev space, and the concept of a weak solution.

The whole problematic is treated in detail in the monograph by K. Rektorys 1980, [2], especially in Chapters 10 and 11, and in Part IV of that book, including corresponding proofs that will not be reproduced here. Thus, the next section will be only a brief survey of these problematics. To make it as clear as possible, we start with the case of functions of one variable, thus with the problematics motivated by boundary value problems in ordinary differential equations cited in Section 3.4.1. Extension to the case of functions of several variables is to be found in Section 3.4.2.

3.4 Generalized Derivatives; The Energetic Space, the Sobolev Space; Generalized Solutions, Weak Solutions

3.4.1 Functions of One Variable

As we know, $C^{(i)}[a, b]$ is the set of all functions continuous in $[a, b]$ simultaneously with their derivatives up to the order i inclusively. By the symbol $C^{(\infty)}[a, b]$ we denote the set of all functions continuous in $[a, b]$ simultaneously with all their derivatives. An important class among these functions is constituted by those that are identically equal to zero in a certain neighborhood of the endpoints $x = a$ and $x = b$ of the given interval (while these neighborhoods may be different for different Figure 3.4.1. functions of that class). A graph of such a function is sketched

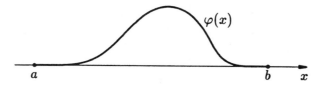

FIGURE 3.4.1

in These functions are often called *functions with compact support in $[a, b]$*, and the set of them is usually denoted by $C_0^{(\infty)}[a, b]$. It can be shown that there is an infinite number of them. (For a mathematically educated reader: $C_0^{(\infty)}[a, b]$ is even dense in $L_2(a,b)$.)

Now, let u be an arbitrary function of $C^{(1)}[a, b]$ (thus u and u' are continuous in $[a, b]$), and let ϕ be an arbitrary function of $C_0^{(\infty)}[a, b]$. Under these assumptions we can obviously integrate by parts, and obtain

$$\int_a^b u'\phi \, dx = [u\phi]_a^b - \int_a^b u\phi' \, dx = - \int_a^b u\phi' \, dx \qquad (3.4.1)$$

since $\phi(a) = 0$ and $\phi(b) = 0$. However, a relation like this may be valid as well for other functions from $L_2(a,b)$, not only for functions from $C^{(1)}[a, b]$. An example follows. Consider function u defined in the following way:

$$u(x) = \begin{cases} -x & \text{for } x \in [-1, 0] \\ x & \text{for } x \in (0, 1] \end{cases} \qquad (3.4.2)$$

(Figure 3.4.2). This function does not belong to $C^{(1)}[-1, 1]$, since it has no derivative at point $x = 0$. In interval $[-1, 0)$, or $(0, 1]$ it has the derivative (at point $x = -1$, or $x = 1$ the right-hand, or left-hand derivative, respectively) equal to -1 or $+1$. If we define function $v(x)$ by value -1 in interval $[-1, 0)$, by 1 in $(0, 1]$

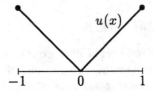

FIGURE 3.4.2

and, for example, by 0 at point $x = 0$ (thus

$$v(x) = \begin{cases} -1 & \text{for } x \in [-1, 0) \\ 0 & \text{for } x = 0 \\ 1 & \text{for } x \in (0, 1] \end{cases} \qquad (3.4.3)$$

see Figure 3.4.3), then we have, for *every* $\phi \in C_0^{(\infty)}[-1, 1]$

$$\int_{-1}^1 v\phi \, dx = \int_{-1}^0 v\phi \, dx + \int_0^1 v\phi \, dx = - \int_{-1}^0 \phi \, dx + \int_0^1 \phi \, dx \qquad (3.4.4)$$

$$= -[x\phi]_{-1}^{0} + \int_{-1}^{0} x\phi' \, dx + [x\phi]_{0}^{1} - \int_{0}^{1} x\phi' \, dx$$

$$= \int_{-1}^{0} x\phi' \, dx - \int_{0}^{1} x\phi' \, dx = - \int_{-1}^{1} u\phi' \, dx$$

(because $\phi(-1) = 0$ and $\phi(1) = 0$). Thus, function v satisfies the same relation as function u' in (3.4.1). So we can consider it as a certain "derivative of function u in a generalized sense." In general, we give this definition:

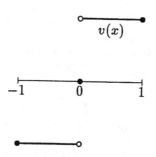

FIGURE 3.4.3

Let $u \in L_2(a,b)$. Let a function $v \in L_2(a,b)$ satisfy the relation

$$\int_{a}^{b} v\phi \, dx = - \int_{a}^{b} u\phi' \, dx \quad \text{for every } \phi \in C_0^{(\infty)}[a, b] \qquad (3.4.5)$$

Then we call v the *generalized derivative* of function u.

Thus, function v considered earlier is a generalized derivative of the function (3.4.2).

If a function u belongs not only to $L_2(a,b)$ but also to $C^{(1)}[a, b]$, then (3.4.5) is satisfied with $v = u'$ (cf. (3.4.1)), and thus u' is not only the derivative of function u in the ordinary sense but also its generalized derivative.

It can be shown that the generalized derivative of a function u is unique. Further, if u has a generalized derivative, then it is a continuous function in $[a, b]$[5]. So the function (3.4.3), for example, cannot have a generalized derivative, because it is not continuous in $[a, b] = [0, 1]$.

Similarly, higher order generalized derivatives (with similar properties) are defined: We say that a function $u \in L_2(a,b)$ has a *generalized derivative* v of the ith

---

[5]For a "mathematical" reader: The function $u(x)$ can be made continuous (even absolutely) in $[a, b]$ when changed, if necessary, on a set of measure zero.

order, if there exists such a function $v \in L_2(a,b)$ that the relation

$$\int_a^b v\phi \, dx = (-1)^i \int_a^b u\phi^{(i)} \, dx \quad \text{holds for every } \phi \in C_0^{(\infty)}[a, b] \qquad (3.4.6)$$

It is usual, in this theory, to denote generalized derivatives of function u by the same symbols $u', u'', \ldots$ as the ordinary ones.

If a function u has the generalized derivative u'', then—as can be shown—it has also the generalized derivative u', while u'' is the generalized derivative of u', and so forth.

Let us denote by M_i the set of all functions $u \in L_2(a,b)$ that have generalized derivatives of the ith order. It follows from what has been said earlier that then every function $u \in M_i$ has also generalized derivatives up to the $(i-1)^{\text{th}}$ order and that these generalized derivatives—including the function u itself—are continuous functions in $[a, b]$.

For our equation

$$Au = f \qquad (3.4.7)$$

with a positive definite operator A of order $2k$, the set M_k, where thus

$$M_k = \{u : u \in L_2(a,b), \ u \text{ has generalized derivatives} \qquad (3.4.8)$$

$$\text{up to the } k^{\text{th}} \text{ order including}\}$$

is of fundamental importance, and the same holds for some of its subsets. What have we in mind follows.

For Equation (3.4.7) linear homogeneous boundary conditions are given at each of the end points $x = a$ and $x = b$ of the interval $[a, b]$. Those that contain derivatives of order $k - 1$ at most (where $2k$ is the order of differential operator A) are called *stable* and the others are *unstable*.

Denote by V_k the subset of set M_k of such functions that satisfy the given stable boundary conditions. Thus

$$V_k = \{u : u \in L_2(a,b), \ u \text{ has generalized derivatives} \qquad (3.4.9)$$

$$\text{up to the } k^{\text{th}} \text{ order including and}$$

$$\text{satisfies the given stable boundary conditions}\}$$

Example 3.4.1

Let us consider the problem

$$-u'' + \left(1 + \cos^2 x\right) u = f(x) \qquad (3.4.10)$$

$$u(0) = 0, \quad u(\pi) = 0 \tag{3.4.11}$$

where $f \in L_2(0, \pi)$ is a given function ($f(x) = \sin x$, e.g., as it was in Example 3.2.1).

Here, operator A, given by

$$Au = -u'' + \left(1 + \cos^2 x\right) u \tag{3.4.12}$$

is an operator of the second order (so that $2k = 2$ and thus $k = 1$), positive definite on its domain of definition

$$D_A = \{u : u \in C^{(2)}[0, \pi], \ u(0) = 0, \ u(\pi) = 0\} \tag{3.4.13}$$

as shown in Example 3.2.2. Both conditions (3.4.11) are stable, because $k = 1$, so that $k - 1 = 0$, and conditions (3.4.11) contain derivatives of order zero only (i.e., the function u itself). By (3.4.9), we have

$$V_1 = \{u : u \in L_2(0, \pi), \ u \text{ has generalized derivatives} \tag{3.4.14}$$

$$\text{of the 1st order, } u(0) = 0, u(\pi) = 0 \}$$

Obviously set V_1 is broader than set D_A, because it also contains more general functions than the functions from $C^{(2)}[0, \pi]$. ☐

Example 3.4.2
Consider the problem (3.3.15) and (3.3.16) from Example 3.3.1

$$\left(EIu''\right)'' + ru = q \tag{3.4.15}$$

$$u(0) = 0, \quad u(l) = 0, \quad u'(0) = 0, \quad u'(l) = 0 \tag{3.4.16}$$

Here, operator A, given by

$$Au = \left(EIu''\right)'' + ru \tag{3.4.17}$$

is an operator of the fourth order ($2k = 4$, $k = 2$), positive definite on its domain of definition

$$D_A = \{u : u \in C^{(4)}[0, l], \ u(0) = 0, \ u(l) = 0, \ u'(0) = 0, \ u'(l) = 0\}$$

(see the quoted example). All the conditions (3.4.16) are stable, since they contain derivatives of the first order at most, and $k - 1 = 1$ here. So

$$V_2 = \{u : u \in L_2(0, l), \ u \text{ has generalized derivatives of the} \qquad (3.4.18)$$

$$\text{2nd order including, } u(0) = 0, \ u(l) = 0, \ u'(0) = 0, \ u'(l) = 0 \}$$

Obviously V_2 is broader than D_A.

If the conditions (3.4.16) were replaced by the conditions

$$u(0) = 0, \quad u(l) = 0, \quad u''(0) = 0, \quad u''(l) = 0 \qquad (3.4.19)$$

(which would correspond to a simply supported bar), then only the two first of them would be stable, the third and the fourth condition being unstable, since $k - 1 = 1$, as stated earlier. Thus, for the problem (3.4.15) and (3.4.19) we would have

$$V_2 = \{u : u \in L_2(0, l), \ u \text{ has generalized derivatives} \qquad (3.4.20)$$

$$\text{of the 2nd order including, } u(0) = 0, \ u(l) = 0\}$$

Obviously, the set (3.4.20) is broader than set D_A, even broader than the set (3.4.18).
☐

Let us stay at the problem (3.4.10) and (3.4.11) of Example 3.4.1, for a moment. Let us introduce, on D_A (see (3.4.13)), the so-called *energetic scalar product* $(u, v)_A$ by

$$(u, v)_A = (Au, v) \qquad (3.4.21)$$

$$u, v \in D_A$$

(Thanks to linearity, symmetry, and positivity of operator A it can be shown—by an easy calculation—that (3.4.21) has all the properties (1.3.8) to (1.3.11) of the scalar product in $L_2(a,b)$, stated in Section 1.3, from which comes its name "scalar product.") Similar to Example 3.2.2 (see (3.2.19)), integrating by parts we get, in our case,

$$(u, v)_A = (Au, v) = \int_0^\pi \left[\left(-u'' + \left(1 + \cos^2 x \right) u \right] v \, dx \qquad (3.4.22)$$

$$= \int_0^\pi u'v' \, dx + \int_0^\pi \left(1 + \cos^2 x \right) uv \, dx$$

$$u, v \in D_A$$

Now, because the functions of set V_1 (see (3.4.14)) have generalized derivatives of the first order (square integrable in $[0, \pi]$ by definition), both integrals on the right-hand side of (3.4.22) have a sense for arbitrary functions $u, v \in V_1$. Thus the energetic scalar product (3.4.22), defined originally for functions of D_A only, can be extended to the whole set V_1 by

$$(u, v)_A = \int_0^\pi u'v' \, dx + \int_0^\pi \left(1 + \cos^2 x\right) uv \, dx \qquad (3.4.23)$$

$$u, v \in V_1$$

Also the functional of energy F, defined originally by

$$Fu = (Au, u) - 2(f, u)$$

only for $u \in D_A$, can be extended by

$$Fu = (u, u)_A - 2(f, u) \qquad (3.4.24)$$

$$= \int_0^\pi u'^2 \, dx + \int_0^\pi \left(1 + \cos^2 x\right) u^2 \, dx - 2\int_0^\pi fu \, dx$$

onto the whole set V_1.

Now, we are very close to the main result of this section.

Let us consider the equation

$$Au = f$$

with an operator A of the $2k^{\text{th}}$ order, positive definite on D_A. Let V_k be the set (3.4.9). Define, on D_A, the so-called *energetic scalar product* $(u, v)_A$ by

$$(u, v)_A = (Au, v)$$

Integrating integrals appearing in (Au, v) by parts, we come (similarly, as in (3.4.22) or Example 3.3.1, by repeated integration in general). So integrals having sense not only for functions from set D_A, but from the broader set V_k as well. So we can extend this scalar product onto the whole V_k. Define further, on V_k, the so-called *energetic norm* $\|u\|_A$ and *energetic distance* $\rho_A(u, v)$ by

$$\|u\|_A = \sqrt{(u, u)_A} \qquad (3.4.25)$$

$$\rho_A(u, v) = \|u - v\|_A \qquad (3.4.26)$$

(We say that by Equation (3.4.26) the *metrics* have been introduced on V_k.) The space obtained this way is called the *energetic space* (or *energy space*) and is denoted by H_A, as usual. (For a mathematical reader: H_A can be shown to be a complete space; thus it is a Hilbert space; D_A is a dense set in H_A.)

The functional of energy, defined originally for functions of D_A only

$$Fu = (Au, u) - 2(f, u) \qquad (3.4.27)$$

can be extended to all functions of H_A (we say briefly onto the whole space H_A) by

$$Fu = (u, v)_A - 2(f, u) \qquad (3.4.28)$$

Now, we can formulate the fundamental theorem of the present theory (always under the assumption of positive definiteness of the operator A on D_A).

THEOREM 3.4.1

The functional (3.4.28) *attains its minimum on* H_A. *The function* $u_0 \in H_A$, *realizing this minimum, is uniquely determined by the right-hand side* $f \in L_2(a,b)$ *of the equation* $Au = f$ (and by operator A, of course, by which the energetic scalar product has been generated).

As announced earlier, the function u_0 is called the *generalized solution* of the given problem (i.e., of the equation $Au = f$ with the given boundary conditions).

Theorem 3.4.1 is a profound theorem. Its proof is not at all easy. We are not going to present it here (and we are not able to do it, because we have not the needed functional–theoretical tools at our disposal). For the proof see, for example, K. Rektorys 1980, [2] Chapters 10 and 11 and Part IV of the monograph.

If we want, we can reformulate Theorem 3.4.1 as follows.

THEOREM 3.4.2

If operator A *is positive definite on* D_A, *then there exists a unique generalized solution* u_0 *of the given problem.*

How to find it, or how to get its sufficiently exact approximation, will be shown in the next section.

It may happen that the generalized solution u_0 belongs to the original domain of definition D_A of operator A. Then it is the solution of the given problem in the "classical sense." This happens if the coefficients of (linear) operator A as well as the right-hand side f of the given equation are "sufficiently smooth." We do not go into detail. If, for example, in the problem (3.4.10) and (3.4.11) we have $f(x) = \sin x$, as it was in Example 3.2.1, then its generalized solution can be shown to belong to D_A; thus this problem has a (unique) classical solution. On the other hand, if the function $f \in L_2(0, \pi)$ is sufficiently discontinuous (e.g., piecewise constant) in the given interval, then this problem has a generalized solution u_0 by Theorem 3.4.1 (or 3.4.2), but (see an analogous consideration in Section 3.3) this generalized solution cannot belong to D_A. This fact shows once more the necessity of extending the original domain D_A. In such a case we know only that u_0 belongs to H_A; thus, it does not need to have even a sufficient number of derivatives that are required by the given equation (two in number in our example). A question then arises, of course, whether the generalized solution of the given problem is a "reasonable" concept. It turns out that it is. In fact, especially elliptical equations are often derived from certain functionals by methods of the so-called variational calculus, and contain then "unnecessarily many" derivatives. In this sense, the generalized solution, based on minimalization of the functional of energy on a proper class of functions (= in a proper space), is a much more natural concept than a classical solution. (Well known is the Hilbert thesis stating, "The laws in physics should be formulated in an integral, not differential form.")

Questions concerning regularity of generalized solutions (or weak solutions, see later) of boundary value problems, especially in partial differential equations, are not at all easy. See, for example, K. Rektorys 1980, [2] Chapter 46.

REMARK 3.4.1   *(the Sobolev space, weak solutions)*

By Theorem 3.4.1 the problem has been solved on how to extend set D_A so that the functional of energy attained its minimum. (By the way, this is an uneasy problem that has occupied outstanding mathematicians for years.) To be able to formulate this theorem, it was necessary for us to introduce some new concepts—generalized derivatives, energetic space H_A, and so forth. Having introduced them, we have the possibility now to introduce further concepts currently used in variational methods of solving boundary value problems, the Sobolev space and the concept of a weak solution. These concepts will not be needed in what follows. A reader who is not interested in them can take this remark for brief information only.

1. The Sobolev space. On set M_k of functions having generalized derivatives up to the kth order (see (3.4.8)), metrics other than those given may be introduced with help of the energetic scalar product. Let us define the so-called *Sobolev scalar product* $(u, v)_k(a, b)$ in brief by

$$(u, v)_{W_2^{(k)}(a,b)} = (u, v) + (u', v') + \cdots + \left(u^{(k)}, v^{(k)}\right) \qquad (3.4.29)$$

where (u, v) is the scalar product in $L_2(a,b)$. For example

$$(u, v)_{W_2^{(1)}(a,b)} = \int_a^b uv \, \mathrm{d}x + \int_a^b u'v' \, \mathrm{d}x$$

On the basis of Equation (3.4.29) let us define the *Sobolev norm,* or *distance* by

$$\|u\|_{W_2^{(k)}(a,b)} = \sqrt{(u, u)_{W_2^{(k)}(a,b)}} \tag{3.4.30}$$

or

$$\rho_{W_2^{(k)}(a,b)}(u, v) = \|v - u\|_{W_2^{(k)}(a,b)} \tag{3.4.31}$$

respectively. Set M_k equipped with these metrics is called the *Sobolev space* $W_2^{(k)}(a, b)$. (Notations $H_k(a, b)$ and $H^k(a, b)$, or briefly H_k and H^k only, if there is no danger of misunderstanding, are also used in the literature.)

Similarly, as subsets V_k of set M_k are used, so are subspaces of space $W_2^{(k)}(a, b)$. Especially by $\overset{\circ}{W}_2^{(k)}(a, b)$ the subspace of $W_2^{(k)}(a, b)$ of such functions is denoted for which

$$u(a) = u'(a) = \cdots = u^{(k-1)}(a) = 0 \tag{3.4.32}$$

$$u(b) = u'(b) = \cdots = u^{(k-1)}(b) = 0$$

2. Weak solutions. Let us show the concept of a weak solution of a boundary-value problem in an example.

Consider the problem (3.4.10) and (3.4.11), that is, the problem

$$-u'' + \left(1 + \cos^2 x\right) u = f(x) \tag{3.4.33}$$

$$u(0) = 0, \quad u(\pi) = 0 \tag{3.4.34}$$

with $f \in L_2(a,b)$. Denote by V the subspace of the Sobolev space $W_2^{(1)}(0, \pi)$ of functions that satisfy the (stable) boundary conditions (3.4.34), that is, let

$$V = \{v : v \in W_2^{(1)}(0, \pi), v(0) = 0, v(\pi) = 0\} \tag{3.4.35}$$

(with the metrics of the space $W_2^{(1)}(0, \pi)$; thus we have $V = \overset{\circ}{W}_2^{(1)}(0, \pi)$, cf. (3.4.32)).

Let the problem (3.4.33) and (3.4.34) have a classical solution u. Choose an arbitrary function $v \in V$, multiply Equation (3.4.33) by this function, and integrate between the limits 0 and π:

$$-\int_0^\pi u''v \, dx + \int_0^\pi \left(1 + \cos^2 x\right) uv \, dx = \int_0^\pi fv \, dx$$

Integrating the first integral by parts and using the fact that $v(0) = 0$ and $v(\pi) = 0$, we obtain (for every $v \in V$, because v has been chosen arbitrarily)

$$\int_0^\pi u'v' \, dx + \int_0^\pi \left(1 + \cos^2 x\right) uv \, dx = \int_0^\pi fv \, dx \qquad (3.4.36)$$

Thus if u is a classical solution of the problem (3.4.33) and (3.4.34), then Equation (3.4.36) is satisfied. We also say that function u satisfies the *integral identity* (3.4.36), because Equation (3.4.36) is fulfilled for *every* $v \in V$.

Now, without regard to whether such a classical solution does or do not exist, we define a function u as a *weak solution of the problem* (3.4.33) and (3.4.34), if $u \in V$ and if Equation (3.4.36) is satisfied for every $v \in V$.

The left-hand side of Equation (3.4.36) is linear in u as well as in v and is called the *bilinear form* (corresponding to operator A and to the conditions (3.4.34). It is denoted by $((u, v))$, as usual. With this notation we can rewrite Equation (3.4.36) in the form

$$((u, v)) = (f, v) \quad \text{for every } v \in V \qquad (3.4.37)$$

In the case of other problems one proceeds similarly. V is the subspace of functions from $W_2^{(k)}(a, b)$ satisfying the given stable boundary conditions. The corresponding bilinear form is generally more complicated.

Let us note that, in our example, the bilinear form $((u, v))$ has exactly the same form as the energetic scalar product $(u, v)_A$; see (3.4.23). However, to the construction of $((u, v))$ no symmetry of operator A was needed. The same holds in the general case. Moreover, no smoothness of coefficients of operator A (e.g., of the function p in Problem 1 of Table 3.1) is required.

To ensure existence of a weak solution of the given problem it is sufficient if the form $((u, v))$ is a so-called V-bounded and V-elliptical one (for the proof, see K. Rektorys 1980, [2] Chapter 33). Roughly speaking, it suffices if operator A has some properties of positivity and if its coefficients are bounded functions (integrable, of course). It suffices, for example, if these coefficients are continuous in $[a, b]$ and if A is positive–definite on D_A.

If the form $((u, v))$ is symmetrical, then it can be proved that a function u is the weak solution of the given problem exactly if it minimizes, on V, the functional

$$Fu = ((u, u)) - 2(f, u)$$

Thus (the boundary conditions being homogeneous, all the time), if the coefficients of operator A are smooth enough and if the problem is symmetrical, the concepts of the weak solution and of the generalized solution are equivalent. (For a mathematical reader: The spaces V and H_A have the same elements and their metrics are equivalent.) In general, the concept of a weak solution is a generalization of that of a generalized solution. Moreover, the first of them makes it possible to include nonhomogeneous boundary conditions in a much more natural way than the second one does. For a detailed theory, see K. Rektorys 1980, [2], especially Chapters 32 to 34.

3.4.2   Functions of Several Variables

Concepts and results introduced in Section 3.4.1 may be carried over almost literally to the case of functions of several variables. For example, we say that a function $u \in L_2(\Omega)$ has a generalized derivative $\partial u / \partial x_1$ if there exists such a function $v \in L_2(\Omega)$ (denoted then by $\partial u / \partial x_1$) that

$$\int_\Omega v\phi \, dx = - \int_\Omega u \frac{\partial \phi}{\partial x_1} \, dx \quad \text{holds for every } \phi \in C_0^{(\infty)}(\Omega)$$

The energetic space H_A is constructed in a quite similar way as in Section 3.4.1); Theorem 3.4.1 or 3.4.2 remains valid. Also the construction of the Sobolev space is similar, and the same holds for the concept of a weak solution.

The only essential difference lies in the fact that if a function $u \in L_2(\Omega)$ has generalized derivatives $\partial u / \partial x_1, \ldots, \partial u / \partial x_N$ with respect to all variables $x_1, \ldots, x_N$, then it does not need to be continuous (not even after having been changed on a set of measure zero, see footnote on page 113) in $\overline{\Omega}$. The consequence of this fact is that boundary conditions cannot then be considered in the ordinary sense, but in the so-called *sense of traces*. For this concept, rather complicated for a nonprofessional mathematician, see K. Rektorys 1980, [2] Chapter 30.

3.5   The Ritz and Galerkin Methods; The Finite-Element Method

The main aim of the preceding section was (under the assumption of positive definiteness of operator A) to construct the so-called energetic space H_A and to present Theorem 3.4.1 according to which the functional of energy Fu, extended from D_A onto H_A, attains its minimum on H_A. The function $u_0 \in H_A$ for which this minimum is realized, is called the generalized solution of the given problem. However, the proof of Theorem 3.4.1 is based on the so-called Riesz theorem, which is an abstract mathematical theorem giving no instructions on how to effectively construct this generalized solution, or at least a sufficiently close approximation to it. This is the task for the so-called *variational methods*. The first of them is the Ritz method.

DEFINITION 3.5.1

We say that the sequence

$$v_1, v_2, \ldots, v_n, \ldots \tag{3.5.1}$$

of linearly independent functions from H_A is a basis in H_A, if it is possible to approximate an arbitrary function $u \in H_A$, with an arbitrary accuracy, by a proper linear combination of the functions (3.5.1). In detail, if to every $u \in H_A$ and to every $\varepsilon > 0$ it is possible to find such an integer j and such numbers $a_1, a_2, \ldots, a_j$ that for the distance ρ_A (see Equation 3.4.26) between the function u and the approximation

$$s_n = a_1 v_1 + \cdots + a_j v_j = \sum_{i=1}^{j} a_i v_i$$

we have

$$\rho_A(u, s_n) < \varepsilon \tag{3.5.2}$$

Similarly, bases in other spaces are defined. For example, in $L_2(0, \pi)$ a well-known basis is

$$v_n = \sqrt{\frac{2}{\pi}} \sin nx \tag{3.5.3}$$

$$n = 1, 2, \ldots$$

(Theorem 1.5.3, Example 1.3.7). Here we have even ($u \in L_2(0, \pi)$ being given)

$$a_i = (u, v_i) \tag{3.5.4}$$

and

$$u = \sum_{i=1}^{\infty} a_i v_i \quad \text{in } L_2(0, \pi) \tag{3.5.5}$$

so that u can be approximated, with an arbitrary accuracy in the metrics of the space $L_2(0, \pi)$, by a sum

$$\sum_{i=1}^{j} a_i v_i$$

(a_i being given by Equation (3.5.4)), if we take j sufficiently large.

A proper choice of a basis in H_A can be seen later.

The *Ritz method* (suggested by the German engineer Walter Ritz) consists of minimizing the functional

$$Fu = (u, u)_A - 2(f, u) \qquad (3.5.6)$$

not on the whole energetic space H_A, but on a certain n-dimensional subspace of it. In more detail, choose an arbitrary, but fixed n and consider the first n functions

$$v_1, v_2, \ldots, v_n \qquad (3.5.7)$$

of the basis (3.5.1). Denote by V_n the n-dimensional subspace of space H_A constituted by all linear combinations

$$\sum_{i=1}^{n} b_i v_i \qquad (3.5.8)$$

of the functions (3.5.7), where thus v_i are fixed and b_i are arbitrary (real) numbers. Among the functions of the form (3.5.8) we want to find such a function

$$u_n = \sum_{i=1}^{n} a_i v_i \qquad (3.5.9)$$

(i.e., such numbers $a_1, \ldots, a_n$) for which the functional (3.5.6) attains its minimum on V_n.

In this way, the "infinite-dimensional" problem (how to find the minimum of F on H_A) is converted into an "n-dimensional" one.

How to proceed concretely, will be shown for case $n = 2$ first, for simplicity. The result will then be easily generalized (see later). Thus, let the functional of energy (3.5.6) be given and let the expression (3.5.1) be a basis in H_A. Let us take the first two terms v_1 and v_2 of this basis. We then have to find, from among all functions of the form

$$b_1 v_1 + b_2 v_2 \qquad (3.5.10)$$

(with v_1 and v_2 fixed), such a function

$$u_2 = a_1 v_1 + a_2 v_2 \qquad (3.5.11)$$

for which the functional (3.5.6) attains its minimum, that is, for which

$$\min_{V_2} F(b_1 v_1 + b_2 v_2) = F(a_1 v_1 + a_2 v_2) \qquad (3.5.12)$$

However, as we have mentioned in the preceding section, the energetic scalar product has all properties (1.3.8) to (1.3.11) of the scalar product in $L_2(a,b)$, so that it can be treated according to the same rules. So we have

$$F(b_1 v_1 + b_2 v_2) = (b_1 v_1 + b_2 v_2, b_1 v_1 + b_2 v_2)_A - 2(f, b_1 v_1 + b_2 v_2) \quad (3.5.13)$$

$$= b_1^2 (v_1, v_1)_A + b_1 b_2 (v_1, v_2)_A + b_2 b_1 (v_2, v_1)_A + b_2^2 (v_2, v_2)_A$$

$$- 2b_1 (f, v_1) - 2b_2 (f, v_2)$$

Because functions f, v_1, v_2 are fixed, all scalar products in (3.5.13) are known numbers. In this way, functional F becomes a quadratic function of variables b_1 and b_2. The necessary (and in our case also sufficient) condition for this functional to attain its minimum at points a_1 and a_2 is that its partial derivatives with respect to b_1 and b_2 be equal to zero at that point

$$\left. \frac{\partial F}{\partial b_1} \right|_{\substack{b_1 = a_1 \\ b_2 = a_2}} = 0, \qquad \left. \frac{\partial F}{\partial b_2} \right|_{\substack{b_1 = a_1 \\ b_2 = a_2}} = 0 \qquad (3.5.14)$$

Thus, differentiating (3.5.13) first with respect to b_1 and then with respect to b_2, and putting $b_1 = a_1, b_2 = a_2$ by (3.5.14), we obtain

$$2a_1 (v_1, v_1)_A + a_2 (v_1, v_2)_A + a_2 (v_2, v_1)_A - 2(f, v_1) = 0 \quad (3.5.15)$$

$$a_1 (v_1, v_2)_A + a_1 (v_2, v_1)_A + 2a_2 (v_2, v_2)_A - 2(f, v_2) = 0$$

However,

$$(v_2, v_1)_A = (v_1, v_2)_A \qquad (3.5.16)$$

by symmetry of the energetic scalar product. So putting (3.5.16) into (3.5.15) and dividing by two, we get finally

$$(v_1, v_1)_A \, a_1 + (v_1, v_2)_A \, a_2 = (f, v_1) \qquad (3.5.17)$$

$$(v_2, v_1)_A \, a_1 + (v_2, v_2)_A \, a_2 = (f, v_2)$$

(with $(v_2, v_1)_A = (v_1, v_2)_A$).

This is the system of equations for the wanted coefficients a_1 and a_2 in (3.5.11). In the case of the space V_n, the system (3.5.17) becomes

$$(v_1, v_1)_A \, a_1 + (v_1, v_2)_A \, a_2 + \ldots + (v_1, v_n)_A \, a_n = (f, v_1) \quad (3.5.18)$$

$$(v_2, v_1)_A \, a_1 + (v_2, v_2)_A \, a_2 + \ldots + (v_2, v_n)_A \, a_n = (f, v_2)$$

$$\ldots\ldots\ldots$$

$$(v_n, v_1)_A \, a_1 + (v_n, v_2)_A \, a_2 + \ldots + (v_n, v_n)_A \, a_n = (f, v_n)$$

If, moreover, the functions (3.5.7) belong to D_A (a frequent case), then the system (3.5.18) can be written in the form

$$(Av_1, v_1) \, a_1 + (Av_1, v_2) \, a_2 + \ldots + (Av_1, v_n) \, a_n = (f, v_1)$$

$$(Av_2, v_1) \, a_1 + (Av_2, v_2) \, a_2 + \ldots + (Av_2, v_n) \, a_n = (f, v_2) \qquad (3.5.19)$$

$$\ldots\ldots\ldots$$

$$(Av_n, v_1) \, a_1 + (Av_n, v_2) \, a_2 + \ldots + (Av_n, v_n) \, a_n = (f, v_n)$$

The system (3.5.18) or (3.5.19) is called the *Ritz system* of equations. Its matrix, the so-called *Ritz matrix*, is symmetrical, obviously. It can be shown that, in consequence of linear independence of the functions $v_1, \ldots, v_n$, its determinant, the so-called *Gram determinant,* is different from zero, and that this system has a unique solution. In this way, we obtain a uniquely determined function

$$u_n = a_1 v_1 + \cdots + a_n v_n \qquad (3.5.20)$$

which minimizes the functional of energy on space V_n. We have to realize that function u_n is not the generalized solution u_0 of the given problem yet, since it does not minimize the functional F on the whole space H_A but it does on its n-dimensional subspace V_n. In this sense it is an approximation to the wanted solution only. We call it the nth *Ritz approximation.* Intuitively it can be expected that for $n \to \infty$ the functions u_n will converge to the wanted generalized solution u_0 of the problem. In fact, we have the theorem that follows.

THEOREM 3.5.1
For $n \to \infty$

$$\lim_{n \to \infty} u_n = u_0$$

in the space H_A as well as in the space $L_2(\Omega)$.

For the proof and for the possibility of getting an effective error estimate (i.e., estimate of the "distance between u_0 and u_n") in space H_A or $L_2(\Omega)$, see K.

Rektorys 1980, [2] Chapter 11. Before giving numerical examples, we show how to choose a basis, at least in the simplest cases.

Choice of a Basis for Equations of the Second Order — The question of how to choose a suitable basis is not easy, in general, because not only theoretical but also practical aspects (numerical stability of the process, etc.) are to be taken into account. Therefore, in K. Rektorys 1980, [2] particular attention has been devoted to this question. Here we are going to present the classical choice of the basis for second-order differential equations with very simple boundary conditions only. For the so-called finite element method see page 133.

1. The equation

$$- (pu')' + ru = f \tag{3.5.21}$$

with boundary conditions

$$u(0) = 0, \quad u(l) = 0 \tag{3.5.22}$$

(a special case of Problem 1 from Table 3.1)

    a. The trigonometric basis

$$v_1 = \sin \frac{\pi x}{l}, \quad v_2 = \sin \frac{2\pi x}{l}, \quad \dots, \quad v_n = \sin \frac{n\pi x}{l}, \quad \dots \tag{3.5.23}$$

    b. The polynomial basis

$$v_1 = g(x), \quad v_2 = xg(x), \quad \dots, \quad v_n = x^{n-1}g(x), \quad \dots \tag{3.5.24}$$

with $g \in C^{(2)}[0, l]$ being a function positive in $(0, l)$ and equal to zero at points $x = 0$ and $x = l$. An example of such a function is

$$g(x) = x(l - x) \tag{3.5.25}$$

2. The equation

$$- \Delta u = f \text{ in } \Omega \tag{3.5.26}$$

with the boundary condition

$$u = 0 \text{ on } \Gamma \tag{3.5.27}$$

(Problem 5 in Table 3.1, considering the case $N = 2$)

    a. If Ω is a rectangle $(0, l_1) \times (0, l_2)$, choose the trigonometric basis

$$v_1 = \sin \frac{\pi x}{l_1} \sin \frac{\pi y}{l_2} \tag{3.5.28}$$

$$v_2 = \sin \frac{2\pi x}{l_1} \sin \frac{\pi y}{l_2}, \quad v_3 = \sin \frac{\pi x}{l_1} \sin \frac{2\pi y}{l_2}$$

$$v_4 = \sin \frac{3\pi x}{l_1} \sin \frac{\pi y}{l_2}, \quad v_5 = \sin \frac{2\pi x}{l_1} \sin \frac{2\pi y}{l_2}, \quad v_6 = \sin \frac{\pi x}{l_1} \sin \frac{3\pi y}{l_2}$$

. . . .

b. In the general case choosing the polynomial basis

$$v_1 = g, \quad v_2 = xg, \quad v_3 = yg ,$$

$$v_4 = x^2 g, \quad v_5 = xyg, \quad v_6 = y^2 g, \quad \ldots \tag{3.5.29}$$

with $g \in C^{(2)}(\overline{\Omega})$ being a positive function in Ω, equal to zero on Γ; if, for example, Ω is a circle with its center at the origin and radius R, choose

$$g = R^2 - x^2 - y^2 \tag{3.5.30}$$

If Ω is a region sketched in Figure 3.5.1 (a rectangle with a circular hole), we choose

$$g = \left(a^2 - x^2\right)\left(b^2 - y^2\right)\left(x^2 + y^2 - R^2\right)$$

REMARK 3.5.1

Concerning the basis (3.5.23), all is in order from the theoretical point of view. From the point of view of numerical stability, especially if a large number of terms are to be used, it is advisable to choose, instead of (3.5.23), the basis

$$v_1 = \sin \frac{\pi x}{l}, \; v_2 = \frac{1}{2} \sin \frac{2\pi x}{l}, \; \ldots, \; v_n = \frac{1}{n} \sin \frac{n\pi x}{l}, \; \ldots \tag{3.5.31}$$

because if we do not do it, then we meet, in the lower right-hand part of the Ritz matrix, large numbers, which have an unfavorable influence on the numerical process.

A similar remark holds for other bases as well. For details see K. Rektorys 1980, [2] Chapters 20 and 25, where the reader also finds instructions on how to choose bases for higher order equations, or for other types of boundary conditions.

Let us present some numerical examples.

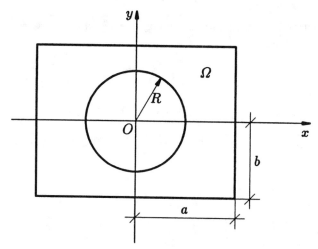

FIGURE 3.5.1

Example 3.5.1
Solve, by the Ritz method, the problem

$$-u'' + \left(1 + \sin^2 x\right) u = 4 \tag{3.5.32}$$

$$u(0) = 0, \quad u(\pi) = 0 \tag{3.5.33}$$

This problem is positive–definite, since it is a special case of Problem 1 of Table 3.1: Here we have $p(x) \equiv 1$, $r(x) = 1 + \sin^2 x$, and $f(x) \equiv 4$, with these functions satisfying assumptions given in that table. (The reader is advised to apply Example 3.2.2 and to prove symmetry and positive definiteness of the corresponding operator, given by

$$Au = -u'' + \left(1 + \sin^2 x\right) u \tag{3.5.34}$$

on its domain of definition

$$D_A = \{u : u \in C^{(2)}[0, \pi], \ u(0) = 0, \ u(\pi) = 0\}) \tag{3.5.35}$$

With regard to the form of function r, we choose the basis (3.5.23) or (3.5.31), so that the computation of the wanted scalar products is as simple as possible. Choose $n = 1$, that is, consider the Ritz approximation in the form

$$u_1 = a_1 v_1 = a_1 \sin x \tag{3.5.36}$$

Because evidently $v_1 \in D_A$, we may use system (3.5.19). For $n = 1$, this system reduces to the single equation

$$(Av_1, v_1) \, a_1 = (f, v_1) \tag{3.5.37}$$

where operator A is given by (3.5.34) and $f(x) \equiv 4$. By (3.5.34) we have

$$Av_1 = -(\sin x)'' + \left(1 + \sin^2 x\right) \sin x = \left(2 + \sin^2 x\right) \sin x$$

so that

$$(Av_1, v_1) = \int_0^\pi \left(2 + \sin^2 x\right) \sin x \cdot \sin x \, dx \tag{3.5.38}$$

$$= \int_0^\pi 2 \sin^2 x \, dx + \int_0^\pi \sin^4 x \, dx = \pi + \left(\frac{1}{4}\pi + \frac{1}{8}\pi\right) = \frac{11}{8}\pi$$

(We apply the well-known formulae $\sin^2 x = \frac{1}{2}(1 - \cos 2x)$, $\cos^2 x = \frac{1}{2}(1 + \cos 2x)$, or use directly tables of integrals.) Further we have

$$(f, v_1) = \int_0^\pi 4 \sin x \, dx = 8$$

Thus by (3.5.37)

$$a_1 = \frac{(f, v_1)}{(Av_1, v_1)} = \frac{64}{11\pi}$$

so that the first Ritz approximation of the solution is

$$u_1 = \frac{64}{11\pi} \sin x \doteq 1.854 \sin x \tag{3.5.39}$$

▯

REMARK 3.5.2
    Especially at points $x_1 = \frac{1}{3}\pi$, and $x_2 = \frac{2}{3}\pi$ we have

$$u_1\left(\frac{1}{3}\pi\right) = u_1\left(\frac{2}{3}\pi\right) \doteq 1.854 \cdot \frac{1}{2}\sqrt{3} \doteq 1.606$$

In Example 1.7.1 we obtained, using the method of finite differences, $z_1 = z_2 \doteq 1.503$. Such a discrepancy could be expected. First, the methods used are different.

Second, when solving the given problem by the method of finite differences, we have chosen a relatively rough division of the given interval into three subintervals only. However, an especially rough approximation has been obtained by the Ritz method, where only one term of the basis has been used.

At this opportunity, let us observe the characteristic feature of the Ritz method: The obtained approximation (3.5.39) fulfills the given boundary conditions exactly and the differential equation, only approximately. If we put (3.5.39) for u into the left-hand side of Equation (3.5.32), we obtain

$$1.854 \sin x + \left(1 + \sin^2 x\right) \cdot 1.854 \sin x = 1.854 \sin x \cdot \left(2 + \sin^2 x\right)$$

and this is a very bad approximation to the right-hand side $f(x) \equiv 4$ of Equation (3.5.32). On the basis of convergence Theorem 3.5.1 it can be expected that, taking a higher number of terms of the basis into consideration, we reach a much better approximation. However, this convergence theorem speaks about convergence in space H_A or L_2 only. Concerning pointwise convergence, the approximation will be bad in the neighborhood of points $x = 0$ and $x = \pi$ even when we take a large number of terms of the basis. In fact, at both these points each term on the left-hand side of Equation (3.5.32) will be equal to zero if we put u_n for u there, since the function u_n will contain only sine terms; and those, as well as their derivatives of the second order, are equal to zero for $x = 0$ and $x = \pi$. The reader is advised to use, instead of function $v_1 = \sin x$, function $v_1 = x(\pi - x) = \pi x - x^2$. Concerning fulfillment of the given equation, an essentially better result can be obtained although in a more lengthy way.

Example 3.5.2
Let us solve, by the Ritz method, the problem

$$-\Delta u = 8 \quad \text{in} \quad \Omega \tag{3.5.40}$$

$$u = 0 \quad \text{on} \quad \Gamma \tag{3.5.41}$$

where Ω is a circle in E_2, with its center at the origin and radius $R = 2$.

The given problem is positive–definite (Problem 5 in Table 3.1). Choose $n = 1$ and

$$v_1 = 4 - x^2 - y^2$$

by (3.5.29) and (3.5.30). We obtain

$$Av_1 = -\Delta v_1 = -\frac{\partial^2 v_1}{\partial x^2} - \frac{\partial^2 v_1}{\partial y^2} = 2 + 2 = 4$$

$$(Av_1, v_1) = \int\int_\Omega 4 \cdot \left(4 - x^2 - y^2\right) dx\, dy$$

Further

$$(f, v_1) = \int\int_\Omega 8 \cdot \left(4 - x^2 - y^2\right) dx\, dy$$

As in the preceding example, the Ritz system is reduced to a single equation

$$(Av_1, v_1)\, a_1 = (f, v_1)$$

that is, to the equation

$$4 \int\int_\Omega \left(4 - x^2 - y^2\right) dx\, dy \cdot a_1 = 8 \int\int_\Omega \left(4 - x^2 - y^2\right) dx\, dy \quad (3.5.42)$$

The integrals in (3.5.42) can be easily computed (using transformation to polar coordinates, see K. Rektorys 1994, [1] Section 14.4); however, it is useless to do it here, since they are equal. (Let us note that by coincidence if the function on the right-hand side is not constant, this case would not occur.) By taking into account that the integral $\int\int_\Omega (4 - x^2 - y^2)\, dx\, dy$ is positive, it immediately follows from (3.5.42)

$$a_1 = 2$$

and consequently

$$u_1 = 2 \left(4 - x^2 - y^2\right)$$

It is of interest that this first Ritz approximation is even the exact solution of the given problem. For further problems see Section 3.7. ⬚

REMARK 3.5.3   *(the Galerkin method)*
   The system (3.5.19) also can be obtained by the so-called *Galerkin method.* This method is based on the requirement that the coefficients a_i in the approximation $u_n = a_1 v_1 + \cdots + a_n v_n$ be such that the function $Au_n - f$ be orthogonal, in space $L_2(\Omega)$, to each of the basic functions $v_1, \ldots, v_n$, that is, that the equations

$$(Au_n - f, v_1) = 0, \quad \ldots, \quad (Au_n - f, v_n) = 0 \quad (3.5.43)$$

be fulfilled. In case operator A is linear, we come, by a simple computation, exactly to the system (3.5.19). If, further, operator A is positive–definite on D_A and if $v_1, \ldots, v_n, \ldots$ ($v_i \in D_A$) is a basis in H_A, then convergence Theorem 3.5.1 remains valid. For details see K. Rektorys 1980, [2] Chapter 14. The requirement in (3.5.43) concerning the coefficients a_i can be formulated as well, of course, when operator A is not positive definite—it does not even need to be symmetrical. Questions connected with solvability of the corresponding system, with convergence of sequence $\{u_n\}$, and so forth then need special investigations. The requirement (3.5.43) can be put down even if operator A is nonlinear. The corresponding system of equations is then nonlinear as well. Its solution is not easy and convergence questions generally are complicated.

REMARK 3.5.4 *(the finite-element method)*
The so-called *finite-element method,* we are going to discuss in the following text, is essentially the Ritz method with a special choice of a basis. Its idea becomes clear from the following example.

Example 3.5.3
Consider the problem

$$-u'' + xu = 2 \tag{3.5.44}$$

$$u(0) = 0, \quad u(1) = 0 \tag{3.5.45}$$

By Table 3.1, this problem is positive–definite. Let us divide interval $[0, 1]$ into $n + 1$ subintervals of the same length $h = 1/(n + 1)$. Construct n functions $v_1, \ldots, v_n$ so that function v_j, $j = 1, \ldots, n$, is continuous in $[0, 1]$, equals zero in intervals $[0, (j-1)h]$ and $[(j+1)h, 1]$, at point $x = jh$ equals 1, and in intervals $[(j-1)h, jh]$ and $[jh, (j+1)h]$ is linear. See Figure 3.5.2, where $n = 4$ has been chosen and the graph of function v_2 has been sketched. Functions of this type are examples of the so-called *spline functions,* the basic property being that they are different from zero only on a "small part" (= on a "small element") of the domain considered. (For other types of spline functions see K. Rektorys 1994, [1] Section 32.9.)
For illustration, let us solve the given problem (3.5.44) and (3.5.45) choosing $n = 2$ (graphs of the corresponding functions v_1 and v_2 are sketched in Figure 3.5.3). Thus let the Ritz approximation be of the form

$$u_2 = a_1 v_1 + a_2 v_2$$

and consider the corresponding Ritz system for determining the coefficients a_1 and a_2. Let us note that it is not possible to use it in the form (3.5.19) here, since the functions v_1 and v_2 do not belong to D_A, not having the second-order derivatives, so that the symbols Av_1 and Av_2 make no sense here. Thus, we have to use the system (3.5.18)

$$(v_1, v_1)_A \, a_1 + (v_1, v_2)_A \, a_2 = (f, v_1) \tag{3.5.46}$$

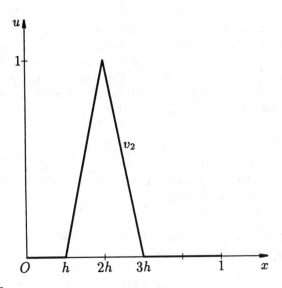

FIGURE 3.5.2

$$(v_2, v_1)_A \, a_1 + (v_2, v_2)_A \, a_2 = (f, v_2)$$

Here

$$(u, v)_A = \int_0^1 u'v' \, dx + \int_0^1 xuv \, dx \qquad (3.5.47)$$

as follows immediately when applying integration by parts to the energetic scalar product

$$(u, v)_A = (Au, v) = \int_0^1 \left(-u'' + xu \right) v \, dx$$

defined originally for functions u and $v \in D_A$ only. (Here, cf. also Equation (3.4.22) ff, it can be seen again how to extend the energetic scalar product from D_A onto the whole energetic space H_A.)

Analytically, functions v_1 and v_2 are given by

$$v_1 = \begin{cases} 3x, \\ 2 - 3x, \\ 0, \end{cases} \quad v_2 = \begin{cases} 0 & \text{for } x \in [0, \frac{1}{3}] \\ 3x - 1 & \text{for } x \in [\frac{1}{3}, \frac{2}{3}] \\ 3 - 3x & \text{for } x \in [\frac{2}{3}, 1] \end{cases} \qquad (3.5.48)$$

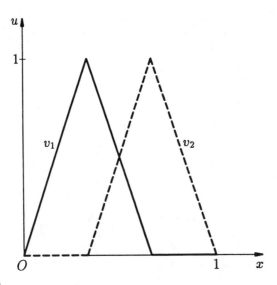

FIGURE 3.5.3

consequently

$$v_1' = \begin{cases} 3, \\ -3, \\ 0, \end{cases} \qquad v_2' = \begin{cases} 0 & \text{for } x \in (0, \tfrac{1}{3}) \\ 3 & \text{for } x \in (\tfrac{1}{3}, \tfrac{2}{3}) \\ -3 & \text{for } x \in (\tfrac{2}{3}, 1) \end{cases} \qquad (3.5.49)$$

Thus by Equation (3.5.47) we have

$$(v_1, v_1)_A = \int_0^1 v_1'^2 \, dx + \int_0^1 x v_1^2 \, dx \qquad (3.5.50)$$

$$= \int_0^{1/3} 3^2 \, dx + \int_{1/3}^{2/3} (-3)^2 \, dx$$

$$+ \int_0^{1/3} x \cdot 9x^2 \, dx + \int_{1/3}^{2/3} x(2 - 3x)^2 \, dx \doteq 6.074$$

and similarly

$$(v_2, v_2)_A = \int_{1/3}^{2/3} 3^2 \, dx + \int_{2/3}^1 (-3)^2 \, dx \qquad (3.5.51)$$

$$+ \int_{1/3}^{2/3} x(3x-1)^2 \, dx + \int_{2/3}^{1} x(3-3x)^2 \, dx \doteq 6.148$$

$$(v_1, v_2)_A = (v_2, v_1)_A \tag{3.5.52}$$

$$= \int_{1/3}^{2/3} 3 \cdot (-3) \, dx + \int_{1/3}^{2/3} x(2-3x)(3x-1) \, dx \doteq -2.972$$

(See also Problem 3.7.4.) Further, we have

$$(f, v_1) = (f, v_2) = \int_0^1 2v_1 \, dx = \frac{2}{3} \tag{3.5.53}$$

In this way, the system (3.5.46) becomes

$$6.074 \, a_1 - 2.972 \, a_2 = 0.667 \tag{3.5.54}$$

$$-2.972 \, a_1 + 6.148 \, a_2 = 0.667$$

from which

$$a_1 \doteq 0.213$$

$$a_2 \doteq 0.212$$

Thus

$$u_2 = 0.213 \, v_1 + 0.212 \, v_2 \tag{3.5.55}$$

The result is represented graphically in Figure 3.5.4. At points $x_1 = \frac{1}{3}$ and $x_2 = \frac{2}{3}$

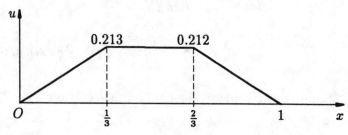

FIGURE 3.5.4

it is in a good accordance with values obtained when solving the same problem (3.5.44) and (3.5.45) by the classical Ritz method (see Problem 3.7.3), or by the finite-difference method (Problem 1.10.3). Also this circumstance shows a close connection between the finite-element method (although it is a variational method) and the method of finite differences. See also Problem 3.7.4. ☐

REMARK 3.5.5

In two-dimensional problems ($N = 2$), the "triangular" spline functions of the type sketched in Figure 3.5.2 are replaced by "pyramidal" functions. See, for example, K. Rektorys 1980, [2] Chapter 42. See also Problem 3.7.16.

The preceding example has hardly convinced the reader of advantages of the finite-element method. On the contrary, the reader probably has come to the conclusion that computation with spline functions is too complicated. Moreover, the result is a nonsmooth function. Advantages of the finite-element method become remarkable only when more complicated problems are solved, especially in partial differential equations. First, this method can be essentially automatized, using special procedures—which we have no possibility of showing here, not even in brief—making it possible to prepare corresponding programs in a relatively simple way. This includes also the case when general regions are considered, where the classical choice of a basis often makes difficulties. Second, the trigonometric basis (3.5.28) can be used in the case of very special regions only; the polynomial basis (3.5.29) can be used for sufficiently general regions. However, functions of that basis are only "slightly" linearly independent, so that the determinant of the Ritz system is nearly zero; consequently, the solution is very sensitive on inaccuracies in computing the needed scalar products as well as on rounding off errors when solving the system. (See also Problem 3.7.14.) This property of "numerical instability" is not encountered when using the finite-element method. Third, by using this method, the matrix of the Ritz system is a bandmatrix, or a matrix of a similar type, which makes the number of operations essentially lower.

These three properties of the finite-element method are so convenient from the practical point of view that this method has become one of the most often used methods when solving elliptical problems (and not only them). See, for example, [4] to [12]. For some other examples see Section 3.7.

3.6 Eigenvalue Problems for Equations of Order $2k$

In Section 1.4, the very simple eigenvalue problem

$$-u'' - \lambda u = 0 \tag{3.6.1}$$

$$u(0) = 0, \quad u(l) = 0 \tag{3.6.2}$$

was treated. (Equation (1.4.1) appears to be multiplied by the number -1 here.) At the same time, solvability of the corresponding nonhomogeneous problem (written here also in a form more convenient for investigation)

$$-u'' - \lambda u = f \tag{3.6.3}$$

$$u(0) = 0, \quad u(l) = 0 \tag{3.6.4}$$

was studied. Basic results were summarized in Theorems 1.5.3 and 1.6.1. The reader is advised to look at these theorems before starting to read the following text.

Main assertions of the quoted theorems can be extended to the much more general case of equations

$$Au - \lambda u = 0 \tag{3.6.5}$$

$$Au - \lambda u = f \tag{3.6.6}$$

with an operator A of order $2k$, positive-definite on its domain of definition D_A, that is, on the set of all its comparison functions. Let us recall that in the definition of comparison functions the given (linear and homogeneous) boundary conditions are already included. This entitles us to speak of Equations (3.6.5) and (3.6.6) as problems (3.6.5) and (3.6.6). Eigenvalues and eigenfunctions of the problem (3.6.5) are defined in a similar way as for the problem (3.6.1) and (3.6.2)[6].

The problem (3.6.1) and (3.6.2), or (3.6.3) and (3.6.4) is a special case of the problem (3.6.5) or (3.6.6), respectively, of course, for A being the positive–definite operator given by

$$Au = -u''$$

on the set of comparison functions

$$D_A = \{u : u \in C^{(2)}[0, l], \ u(0) = 0, \ u(l) = 0\}$$

One of the main differences between the classical problem (3.6.1) and (3.6.2) and the more general problem (3.6.5) consists of the fact that in the case of partial differential equations more than one linearly independent eigenfunction may

---

[6]Exact definitions to which some new concepts are to be introduced, since the solution should be investigated in a generalized sense, can be found in K. Rektorys 1980, [2] Chapter 39.

correspond to a single eigenvalue λ. However, it can be shown that only a finite number of linearly independent eigenfunctions may exist corresponding to such a λ. By using then the so-called orthogonalization process, these eigenfunctions can be orthonormalized. (In the sense of scalar product in the space H_A, thus in the sense

$$\left(v_i, v_j\right)_A = 0 \quad \text{for } i \neq j \tag{3.6.7}$$

$$\left(v_j, v_j\right)_A = 1)$$

For the general case of problems (3.6.5) and (3.6.6) with a positive–definite operator A, the theory is rather complicated. The reader is referred to the monograph K. Rektorys 1980, [2] Chapter 39. Here we state—without proof—only the main result (see, however, Problem 3.7.15).

THEOREM 3.6.1
Let A be a positive–definite operator on D_A. Then

1. There exists a countable set of eigenvalues of the problem (3.6.5), each of them being positive. To each of them there exist infinitely many eigenfunctions, however, only a finite number of linearly independent ones.

2. Eigenfunctions corresponding to different eigenvalues are orthogonal in the space H_A (i.e., in the sense (3.6.7) as well as in space $L_2(\Omega)$).

3. Orthonormal system of eigenfunctions of the problem (3.6.5) constitutes a basis in the space H_A as well as in the space $L_2(\Omega)$.

4. Let λ not be an eigenvalue of the problem (3.6.5). Then the problem (3.6.6) has a unique solution (a generalized one) for every $f \in L_2(\Omega)$.

5. Let λ be an eigenvalue of the problem (3.6.5). Then the problem (3.6.6) is solvable (not uniquely, at the same time) if and only if the function f is orthogonal, in $L_2(\Omega)$, to every eigenfunction corresponding to this λ.

REMARK 3.6.1
Theorem 3.6.1 can be generalized, further, to the case of problems of the form

$$Au - \lambda Bu = 0$$

where A and B are positive–definite operators and B is of a lower order than A. For details see K. Rektorys 1980, [2] Chapter 39. From this generalized theorem—in addition to others—assertions follow concerning the problems (1.6.30), (1.6.31), (1.6.32), and (1.6.33) in Chapter 1. (It is sufficient to realize that operator B given by $Bu = gu$ on the set of continuous functions is positive–definite when g is a positive continuous function.)

REMARK 3.6.2   *(eigenvalue problem for the Laplace operator)*

In some simple cases it is easy to find eigenvalues and eigenfunctions of the problem (3.6.5) explicitly. For example, eigenfunctions of the problem

$$-\Delta u - \lambda u = 0 \text{ in } \Omega \tag{3.6.8}$$

$$u = 0 \text{ on } \Gamma \tag{3.6.9}$$

where Ω is the rectangle $(0, l_1) \times (0, l_2)$, are the functions (3.5.28). Corresponding eigenvalues are, as the reader easily can verify

$$\lambda_1 = \pi^2 \left(\frac{1}{l_1^2} + \frac{1}{l_2^2} \right), \lambda_2 = \pi^2 \left(\frac{4}{l_1^2} + \frac{1}{l_2^2} \right), \tag{3.6.10}$$

$$\lambda_3 = \pi^2 \left(\frac{1}{l_1^2} + \frac{4}{l_2^2} \right), \lambda_4 = \pi^2 \left(\frac{9}{l_1^2} + \frac{1}{l_2^2} \right), \ldots$$

Here, it can be well seen that several linearly independent eigenfunctions may correspond to an eigenvalue λ. If $l_1 = l_2 = l$, that is, if Ω is a square, then

$$\lambda_2 = \lambda_3 = \frac{5\pi^2}{l^2}$$

and to this eigenvalue two linearly independent eigenfunctions correspond

$$\sin \frac{2\pi x}{l} \sin \frac{\pi y}{l}$$

$$\sin \frac{\pi x}{l} \sin \frac{2\pi y}{l} \quad \blacksquare$$

In general, eigenvalues are to be found using approximate methods. For details the reader is referred to K. Rektorys 1980, [2] Chapter 40. Special attention is devoted to the approximation of the first eigenvalue there, which in physical and engineering problems—especially in those concerning stability—is of particular importance. See also Chapter 41 of the quoted book with numerical examples. See also K. Rektorys 1994, [1] Section 17.17.

3.7   Problems 3.7.1 to 3.7.16

3.7.1—Note that

1. The region Ω sketched in Figure 2.2.1 is a bounded region with a Lipschitz boundary.

2. The region Ω_1 shown in Figure 3.7.1 is not a region with a Lipschitz boundary. Why?

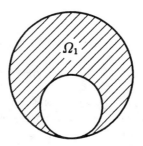

FIGURE 3.7.1

3.7.2—Show that for the problem

$$-\Delta u = f \text{ in } \Omega$$

$$u = 0 \text{ on } \Gamma$$

where Ω is the square $(0, \pi) \times (0, \pi)$, the function

1.
$$u = \sin x \sin y$$

is, and

2.
$$u = \sin x \sinh y$$

is not a comparison function.

3.7.3—In Section 1.10 the problem

$$-u'' + xu = 2 \tag{3.7.1}$$

$$u(0) = 0 \qquad\qquad (3.7.2)$$

$$u(1) = 0$$

(Problem 1.10.3 with Equation (3.7.1) written in the form

$$u'' - xu = -2)$$

was solved approximately by the finite-difference method with step $h = \frac{1}{3}$. As a result one obtained (approximate) values

$$z_1 \doteq 0.212$$

or

$$z_2 \doteq 0.209$$

of the solution at points $x_1 = \frac{1}{3}$ or $x_2 = \frac{2}{3}$, respectively. Solve the same problem by the Ritz method, assuming the first Ritz approximation in the form

$$u_1 = a_1 \left(x - x^2 \right)$$

$[u_1 = \frac{20}{21} x(1 - x); \; u_1(\frac{1}{3}) = u_1(\frac{2}{3}) \doteq 0.211.]$

3.7.4*—In Section 3.5 the same problem has been solved by the finite-element method, choosing two spline functions v_1 and v_2 only. Show how to solve it, by the same method, with n spline functions $v_1, \ldots, v_n$, where n is an arbitrary positive integer $(n > 1)$.

Hint: We have to determine coefficients $(v_i, v_j)_A$ of $a_1, \ldots, a_n$ in system (3.5.18). However,

$$v_i(x) = \begin{cases} \frac{1}{h}[x - (i - 1)h] & \text{for } x \in [(i - 1)h, ih] \\ \frac{1}{h}[(i + 1)h - x] & \text{for } x \in [ih, (i + 1)h] \end{cases}$$

$i = 1, \ldots, n, \; h = 1/(n + 1)$ (Figure 3.7.2), and $v_i(x) \equiv 0$ elsewhere in $[0, 1]$. Thus by (3.5.47)

$$(v_i, v_i)_A = \int_{(i-1)h}^{(i+1)h} v_i' v_i' \, dx + \int_{(i-1)h}^{(i+1)h} x v_i v_i \, dx \qquad (3.7.3)$$

$$\approx \int_{(i-1)h}^{ih} \frac{1}{h^2} \, dx + \int_{ih}^{(i+1)h} \left(-\frac{1}{h} \right) \left(-\frac{1}{h} \right) dx$$

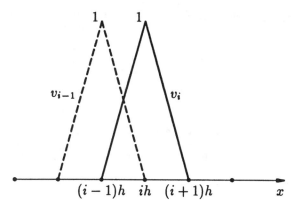

FIGURE 3.7.2

$$+ \frac{ih}{h^2} \left\{ \int_{(i-1)h}^{ih} [x - (i-1)h]^2 \, dx + \int_{ih}^{(i+1)h} [(i+1)h - x]^2 \, dx \right\}$$

$$= \frac{2}{h} + \frac{i}{h} \left\{ \int_0^h z^2 \, dz - \int_h^0 t^2 \, dt \right\} = \frac{2}{h} + \frac{2i}{3} h^2$$

(where in the second integral on the right-hand side of Equation (3.7.3) x was replaced by the "mean value" ih and then the substitutions $x - (i-1)h = z$ or $(i+1)h - x = t$ were applied; simplifications like this are often used in the finite-element method)

$$(v_{i-1}, v_i)_A = \int_{(i-1)h}^{ih} v_{i-1}' v_i' \, dx + \int_{(i-1)h}^{ih} x v_{i-1} v_i \, dx \qquad (3.7.4)$$

$$\approx \int_{(i-1)h}^{ih} \left(-\frac{1}{h}\right) \frac{1}{h} \, dx$$

$$+ \frac{\left(i - \frac{1}{2}\right) h}{h^2} \int_{(i-1)h}^{ih} (ih - x)[x - (i-1)h] \, dx$$

$$= -\frac{1}{h} + \frac{i - \frac{1}{2}}{h} \cdot \frac{h^3}{6} = -\frac{1}{h} + \frac{i - \frac{1}{2}}{6} h^2$$

Especially, in Example 3.5.3 we had $n = 2$ and thus $h = 1/(n+1) = \frac{1}{3}$; and for

$(v_1, v_1)_A$, $(v_2, v_2)_A$, and $(v_1, v_2)_A$ we obtained (3.5.50), (3.5.51), and (3.5.52). (Note that $i = 2$ in (3.7.4).)

3.7.5—Show how to convert the problem

$$-u'' + xu = 4, \quad u(0) = 5, \quad u(1) = -3$$

with nonhomogeneous boundary conditions into that with homogeneous ones.

Hint: Substitution $u = v + 5 - 8x$ turns the given problem into that of

$$-v'' + xv = 4 - x(5 - 8x) = 8x^2 - 5x + 4$$

$$v(0) = 0, \quad v(1) = 0$$

3.7.6—Find the second Ritz approximation u_2 to the problem

$$-u'' + (4 + x)u = 3$$

$$u(0) = 0, \quad u(1) = 0$$

Assume this approximation in the form

$$u_2 = a_1 v_1 + a_2 v_2 = a_1 \left(x - x^2 \right) + a_2 x \left(x - x^2 \right)$$

$$= a_1 \underbrace{\left(x - x^2 \right)}_{v_1} + a_2 \underbrace{\left(x^2 - x^3 \right)}_{v_2}$$

$[\ (Av_1, v_1) = \frac{29}{60},\ (Av_2, v_2) = \frac{149}{840},\ (Av_1, v_2) = (Av_2, v_1) = \frac{102}{420},$

$$(v_1, f) = \frac{1}{2}$$

$$(v_2, f) = \frac{1}{4}$$

$$406\,a_1 + 204\,a_2 = 420$$

$$204\,a_1 + 149\,a_2 = 210$$

$$a_1 \doteq 1.046$$

$$a_2 \doteq -0.023$$

$$u_2 = 1.046 \left(x - x^2 \right) - 0.023 \left(x^2 - x^3 \right) \Big]$$

In the following two problems the reader has the possibility of comparing results of solving the same differential equation with boundary conditions of different types by different methods.

3.7.7—Solve the problem

$$-u'' = 1$$

$$u(0) = 0, \quad u(1) = 0$$

1. Exactly
2. By the finite-difference method with step $h = \frac{1}{3}$
3. By the (classical) Ritz method, assuming the first Ritz approximation in the form $u_1 = a_1(x - x^2)$
4. By the finite-element method with the functions v_1 and v_2 of the same form as in Example 3.5.3

Compare the results obtained by different methods at points $x_1 = \frac{1}{3}$ and $x_2 = \frac{2}{3}$.

[1. $u = \frac{1}{2}(x - x^2)$; $u(x_1) = \frac{1}{9}$, $u(x_2) = \frac{1}{9}$

2. $z_1 = \frac{1}{9}$, $z_2 = \frac{1}{9}$

3. $u = \frac{1}{2}(x - x^2)$; $u(x_1) = \frac{1}{9}$, $u(x_2) = \frac{1}{9}$

4. $u_2 = a_1 v_1 + a_2 v_2 = \frac{1}{9} v_1 + \frac{1}{9} v_2$

Thus, in this (very simple) case all methods lead to the same result.]

3.7.8—Solve the problem

$$-u'' = 1$$

$$u(0) = 0, \quad u'(1) = 0$$

by the same methods (1 to 4) as in Problem 3.7.7 and compare the results at the same points $x_1 = \frac{1}{3}$ and $x_2 = \frac{2}{3}$. Assume the first Ritz approximation u_1 in the form $u_1 = a_1(x - \frac{1}{2}x^2)$; in the case of the finite-difference and finite-element methods use the schemes shown in Figures 3.7.3 and 3.7.4 characterizing the condition $u'(1) = 0$.

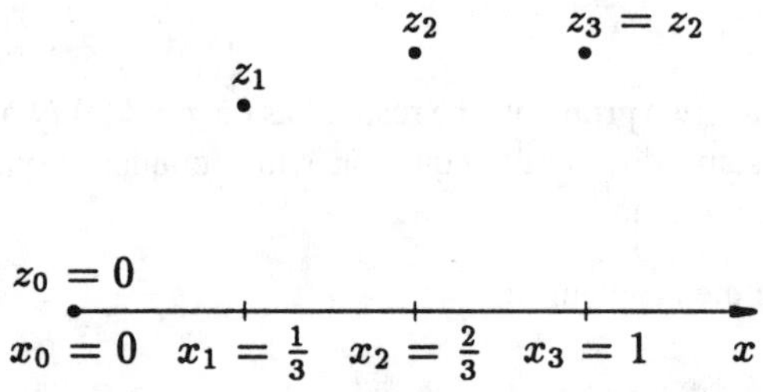

$$z_2 \qquad z_3 = z_2$$
$$z_1$$

$$z_0 = 0$$
$$x_0 = 0 \quad x_1 = \tfrac{1}{3} \quad x_2 = \tfrac{2}{3} \quad x_3 = 1 \qquad x$$

FIGURE 3.7.3

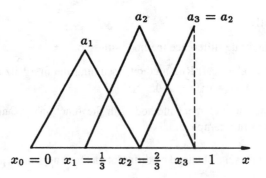

$$x_0 = 0 \quad x_1 = \tfrac{1}{3} \quad x_2 = \tfrac{2}{3} \quad x_3 = 1 \qquad x$$

FIGURE 3.7.4

[1. $u = x - \frac{1}{2}x^2$; $u(x_1) = \frac{5}{18}$, $u(x_2) = \frac{8}{18}$, $u(1) = \frac{9}{18}$

2. $z_1 = \frac{2}{9}$, $z_2 = \frac{3}{9}$, $z_3 = z_2 = \frac{3}{9}$

3. $u = x - \frac{1}{2}x^2$; $u(x_1) = \frac{5}{18}$, $u(x_2) = \frac{8}{18}$, $u(1) = \frac{9}{18}$

4. $u_3 = a_1 v_1 + a_2 v_2 + a_2 v_3$;

$$a_1 (v_1, v_1)_A + a_2 (v_1, v_2)_A + a_2 (v_1, v_3)_A = (1, v_1)$$

$$a_1 (v_2, v_1)_A + a_2 (v_2, v_2)_A + a_2 (v_2, v_3)_A = (1, v_2)$$

$$6a_1 - 3a_2 = \frac{1}{3}$$

$$-3a_1 + 6a_2 - 3a_2 = \frac{1}{3}$$

$$a_1 = \tfrac{2}{9}, \quad a_2 = \tfrac{3}{9}; \quad u_3(\tfrac{1}{3}) = \tfrac{2}{9}, \quad u_3(\tfrac{2}{3}) = u_3(1) = \tfrac{3}{9}.]$$

3.7.9—Convert the problem

$$-\Delta u \equiv -\left(\frac{\partial^2 u}{\partial x^2} + \frac{\partial^2 u}{\partial y^2} \right) = 0 \text{ in } \Omega = (0, 1) \times (0, 1)$$

$$u = 0 \qquad \text{for } y = 0,\ 0 \leqq x \leqq 1$$

$$\text{for } x = 0,\ 0 \leqq y \leqq 1$$

$$\text{for } x = 1,\ 0 \leqq y \leqq 1$$

$$u = \sin \pi x \qquad \text{for } y = 1,\ 0 \leqq x \leqq 1$$

into a problem with zero boundary conditions.

Hint: Put $u = v + y \sin \pi x$; then v satisfies

$$-\Delta v = -\pi^2 y \sin \pi x \text{ in } \Omega$$

$$v = 0 \text{ on } \Gamma$$

3.7.10—By using the Ritz method, solve approximately the problem

$$-\Delta u = 1 \text{ in } \Omega$$

$$u = 0 \text{ on } \Gamma$$

where Ω is the square $(0, \pi) \times (0, \pi)$. Choose the Ritz approximation in the form

$$u_1 = a_1 \underbrace{\sin x \sin y}_{v_1}$$

$[u_1 = 8\pi^{-2} \sin x \sin y.]$

3.7.11—By using the same method, solve the problem

$$-\Delta u \equiv -\left(\frac{\partial^2 u}{\partial x^2} + \frac{\partial^2 u}{\partial y^2}\right) = x^2 + 4y^2 \text{ in } E$$

$$u = 0 \text{ on } \Gamma$$

where E is the interior of the ellipse Γ with semiaxis 2 or 1, in the (positive) x- or y-axis, respectively. Assume the Ritz approximation in the form

$$u_1 = a_1 \underbrace{\left(1 - \frac{1}{4}x^2 - y^2\right)}_{v_1}$$

Hint: When evaluating double integrals over E, use the substitution

$$x = 2r \cos\phi, \quad y = r \sin\phi$$

(see, e.g., K. Rektorys 1994, [1] Remark 14.4.2). You get

$$(Av_1, v_1) = \tfrac{5}{2}\pi, \quad (f, v_1) = \tfrac{4}{3}\pi \implies a_1 = \tfrac{8}{15}$$

$$u_1 = \tfrac{8}{15}\left(1 - \tfrac{1}{4}x^2 - y^2\right)$$

3.7.12*—Consider a horizontal rectangular plate, clamped and loaded vertically. The problem of finding its vertical deflection consists of solving the equation

$$\Delta^2 u \equiv \frac{\partial^4 u}{\partial x^4} + 2\frac{\partial^4 u}{\partial x^2 \partial y^2} + \frac{\partial^4 u}{\partial y^4} = f(x, y) \text{ in } \Omega = (0, a) \times (0, b) \quad (3.7.5)$$

with boundary conditions

$$u = 0 \qquad\qquad\qquad\qquad (3.7.6)$$

$$\frac{\partial u}{\partial \nu} = 0 \text{ on } \Gamma$$

($\partial u/\partial v$ is the outward-normal derivative). Write the domain of definition of operator A and show that this operator is symmetrical and positive on D_A.

Hint: Take into consideration that the conditions (3.7.6) imply

$$\frac{\partial u}{\partial x} = 0$$

$$\frac{\partial u}{\partial y} = 0 \text{ on } \Gamma$$

and use Equation (3.2.31) twice. (It is advantageous here to write the derivative $\partial^4 u/(\partial x^2 \partial y^2)$ in the form $\partial^4 u/(\partial x\, \partial y\, \partial x\, \partial y)$.)

3.7.13—By using results obtained in Problem 3.7.12, show that the corresponding functional of energy is

$$Fu = \int\int_\Omega \left[\left(\frac{\partial^2 u}{\partial x^2}\right)^2 + 2\left(\frac{\partial^2 u}{\partial x \partial y}\right)^2 + \left(\frac{\partial^2 u}{\partial y^2}\right)^2 \right] dx\, dy - 2\int\int_\Omega fu\, dx\, dy$$

3.7.14*—(A plate equation on an annulus, solution by the Ritz method) Consider the problem

$$\Delta^2 u = 1 \text{ in } \Omega, \quad u = 0, \quad \frac{\partial u}{\partial v} = 0 \text{ on } \Gamma$$

where Ω is an annulus with the center at the origin and inner, or outer radius $R_1 = 1$ or $R_2 = 2$, respectively (displacement of a horizontal plate of the form of an annulus, clamped, with a constant vertical loading). Introduce polar coordinates ρ and ϕ and solve the problem, by using the Ritz method with

$$u_2 = a_1 v_1 + a_2 v_2 = a_1 \left(\rho^2 - 1\right)^2 \left(4 - \rho^2\right)^2 + a_2 \rho^2 \left(\rho^2 - 1\right)^2 \left(4 - \rho^2\right)^2$$

Hint: The solution u may be assumed independent of ϕ. Then (cf. K. Rektorys 1994, [1] Section 12.11)

$$\Delta^2 u = \Delta(\Delta u) = \left(\frac{d^2}{dr^2} + \frac{1}{r}\frac{d}{dr}\right)\left(\frac{d^2 u}{dr^2} + \frac{1}{r}\frac{du}{dr}\right)$$

$$= \frac{d^4 u}{dr^4} + \frac{2}{r}\frac{d^3 u}{dr^3} - \frac{1}{r^2}\frac{d^2 u}{dr^2} + \frac{1}{r}\frac{du}{dr}$$

$$\left(\Delta^2 v_1, v_1\right) = \int_1^2 \int_0^{2\pi} \left(2304\rho^4 - 5760\rho^2 + 2112\right)\left(\rho^2 - 1\right)^2\left(4 - \rho^2\right)^2 \rho\, d\rho\, d\phi$$

$$= 72\,588.95$$

and, similarly

$$\left(\Delta^2 v_1, v_2\right) = 228\,434.71, \quad \left(\Delta^2 v_2, v_2\right) = 805\,459.24$$

$$(f, v_1) = 25.447, \quad (f, v_2) = 63.617$$

$$a_1 \doteq 0.000\,948\,5, \quad a_2 \doteq -0.000\,190\,0$$

$$u_2 = \left(0.000\,948\,5 - 0.000\,190\,0\rho^2\right)\left(\rho^2 - 1\right)^2\left(4 - \rho^2\right)^2$$

(The coefficients $(\Delta^2 v_1, v_1), \ldots$ are to be computed with a high accuracy, because the vectors

$$(72\,588.95, \ 228\,434.71)$$

$$(228\,434.71, \ 805\,459.24)$$

are "almost linearly dependent" (cf. Remark 3.5.5). Something like this cannot happen when using the finite-element method.)

3.7.15*—Consider the equation

$$Au - \lambda u = 0 \tag{3.7.7}$$

with linear homogeneous boundary conditions. (See Section 3.6.) Let A be a positive (and consequently symmetrical) operator on its domain of definition D_A. Prove that

1. Eigenvalues of Equation (3.7.7) may be only positive.

2. Eigenfunctions v_i and v_j, corresponding to different eigenvalues λ_i and λ_j, are orthogonal in $L_2(\Omega)$.

Hint: Follow the ideas of proofs of Theorems 1.5.1 and 1.5.2.

3.7.16*—In Remark 3.5.5 we mentioned that equations of the second-order triangular spline functions, applied in the finite-element method for solving one-dimensional problems, are to be replaced by pyramidal ones in two-dimensional cases. By making an analogue to Problem 3.7.4, show how to solve, using these pyramidal functions, the problem

$$-\Delta u = f(x, y) \ \text{ in } \ \Omega$$

$$u = 0 \ \text{ on } \ \Gamma$$

where Ω is a rectangle shown in Figure 3.7.5.

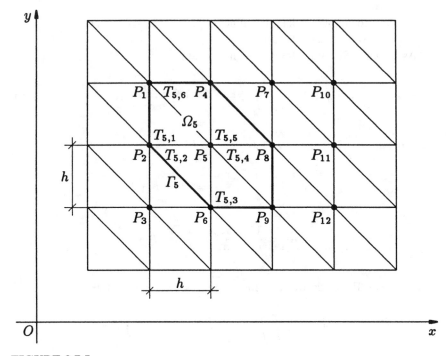

FIGURE 3.7.5

Hint: Divide the rectangle Ω into squares of side-length h as shown in Figure 3.7.5. Divide each of these squares in the way shown in the same figure. You get a triangular mesh with interior mesh points $P_1, \ldots, P_{12}$ in our case. Consider the fifth mesh point $P_5(x_5, y_5)$, for example, and denote by $T_{5,1}, \ldots, T_{5,6}$ triangles with a common vertex at this point. They form a region together—denote it by Ω_5—with the boundary Γ_5 shown in the figure.

Construct the pyramidal spline function v_5, corresponding to point P_5, in the following way: It is continuous in $\overline{\Omega}_5$, assumes the value 1 at point P_5, is linear over each of the triangles $T_{5,1}, \ldots, T_{5,6}$ and equals zero on Γ_5. Outside of Ω_5 it is identically zero. Equations of parts of v_5 on individual triangles $T_{5,1}, \ldots, T_{5,6}$ in the Cartesian system $(O; x, y, v)$ can be easily found. For example

$$v_5 = 1 + \frac{x - x_5}{h} \quad \text{on } T_{5,1} \tag{3.7.8}$$

$$v_5 = 1 + \frac{x - x_5}{h} + \frac{y - y_5}{h} \quad \text{on } T_{5,2} \tag{3.7.9}$$

and so forth.

Similarly, other pyramidal spline functions $v_1, \ldots, v_4, v_6, \ldots, v_{12}$, corresponding to other vertices, are constructed.

As shown in Section 3.2, we have, in our case

$$(v_i, v_j)_A = \int \int_\Omega \left(\frac{\partial v_i}{\partial x} \frac{\partial v_j}{\partial x} + \frac{\partial v_i}{\partial y} \frac{\partial v_j}{\partial y} \right) dx \, dy \tag{3.7.10}$$

By Equations (3.7.8) and (3.7.9) we get

$$\frac{\partial v_5}{\partial x} = \frac{1}{h}, \quad \frac{\partial v_5}{\partial y} = 0 \quad \text{on } T_{5,1}$$

$$\frac{\partial v_5}{\partial x} = \frac{1}{h}, \quad \frac{\partial v_5}{\partial y} = \frac{1}{h} \quad \text{on } T_{5,2}$$

and so forth. Evidently we have, for function v_5 and its derivatives

$$\int \int_\Omega = \int \int_{\Omega_5} = \int \int_{T_{5,1}} + \cdots + \int \int_{T_{5,6}}$$

Thus

$$(v_5, v_5)_A = \int \int_{T_{5,1}} \left(\frac{\partial v_5}{\partial x} \frac{\partial v_5}{\partial x} + \frac{\partial v_5}{\partial y} \frac{\partial v_5}{\partial y} \right) dx \, dy + \cdots$$

$$+ \int \int_{T_{5,6}} \left(\frac{\partial v_5}{\partial x} \frac{\partial v_5}{\partial x} + \frac{\partial v_5}{\partial y} \frac{\partial v_5}{\partial y} \right) dx \, dy$$

$$= \frac{h^2}{2} \left[\left(\frac{1}{h} \cdot \frac{1}{h} + 0 \cdot 0 \right) + \left(\frac{1}{h} \cdot \frac{1}{h} + \frac{1}{h} \cdot \frac{1}{h} \right) + \left(0 \cdot 0 + \frac{1}{h} \cdot \frac{1}{h} \right) \right.$$

$$+ \left(\frac{1}{h} \cdot \frac{1}{h} + 0 \cdot 0\right) + \left(\frac{1}{h} \cdot \frac{1}{h} + \frac{1}{h} \cdot \frac{1}{h}\right) + \left(0 \cdot 0 + \frac{1}{h} \cdot \frac{1}{h}\right)\Bigg] = 4$$

$$(v_5, v_1)_A = \int\int_{T_{5,6}} \left(\frac{\partial v_5}{\partial x}\frac{\partial v_1}{\partial x} + \frac{\partial v_5}{\partial y}\frac{\partial v_1}{\partial y}\right) dx\, dy$$

$$+ \int\int_{T_{5,1}} \left(\frac{\partial v_5}{\partial x}\frac{\partial v_1}{\partial x} + \frac{\partial v_5}{\partial y}\frac{\partial v_1}{\partial y}\right) dx\, dy$$

$$= \frac{h^2}{2}\left\{\left[0 \cdot \left(-\frac{1}{h}\right) + \left(-\frac{1}{h}\right) \cdot 0\right] + \left[\frac{1}{h} \cdot 0 + 0 \cdot \frac{1}{h}\right]\right\} = 0$$

$$(v_5, v_2)_A = \frac{h^2}{2} \cdot \left(-\frac{2}{h^2}\right) = -1$$

$$(v_5, v_6)_A = -1$$

$$(v_5, v_9)_A = 0$$

$$(v_5, v_8)_A = -1$$

$$(v_5, v_4)_A = -1$$

Thus the fifth equation of the system (3.5.18) reads

$$-a_2 - a_4 + 4a_5 - a_6 - a_8 = \int\int_{\Omega_5} f v_5 \, dx\, dy$$

This result suggests that the Ritz system of equations will be a system with a sparse matrix. (This property is typical for the finite-element method.)

The integrals

$$\int\int_{\Omega_i} f v_i \, dx\, dy$$

are computed using simple quadrature formulae, as usual.

Here, only main ideas of the finite-element method have been shown. The whole theory, as well as numerical aspects, deserves special books. See especially [4] to [12].

Chapter 4

The Finite-Difference Method for Partial Differential Equations; The Method of Discretization in Time (the Rothe Method)

4.1 The Finite-Difference Method (the Method of Finite Differences, the Net Method) for Partial Differential Equations

In Section 1.7, the reader was introduced to the finite-difference method for solving boundary value problems in ordinary differential equations. We have given the difference formulae (1.7.1) and (1.7.2) and presented Example 1.7.1, where the basic idea and the practice of the method were demonstrated.

The basic idea of this method for *partial* differential equations is similar to that just mentioned. In the region considered, the so-called *net* (or *mesh*, or *grid*) is constructed, and at the so-called *mesh points* (*net points*), derivatives of the wanted solution are replaced by corresponding difference quotients. Also here we show, in a simple example, how to proceed when using this method.

Example 4.1.1

Let us consider the problem

$$-\Delta u = f \text{ in } \Omega \qquad (4.1.1)$$

$$u = g \text{ on } \Gamma \qquad (4.1.2)$$

on the rectangle $\Omega = (0, l_1) \times (0, l_2)$ with the boundary Γ. Here Δ is the Laplace operator. In our case we thus have

$$\Delta u = \frac{\partial^2 u}{\partial x^2} + \frac{\partial^2 u}{\partial y^2} \tag{4.1.3}$$

with f a given function on Ω and g a given function on Γ. (We thus solve the Dirichlet problem for a Poisson equation, see Section 2.2.)

Let us construct the net on Ω in the following way, for example: The interval $[0, l_1]$ will be divided into four subintervals of length $h = l_1/4$, the interval $[0, l_2]$ into three subintervals of length $k = l_2/3$ (Figure 4.1.1). This net will contain six so-called inner (or interior) mesh points. Let us assign them the numbers $1, 2, 3, 4, 5$, and 6 and denote by $u_1, \ldots, u_6$ the wanted values of the so-called net solution (or finite-difference solution, or finite-difference approximation) at these points (thus of the approximate solution obtained by the method of finite differences—see later). At 14 so-called boundary mesh points on Γ, the values of u are known—they are given by the values $g_1, \ldots, g_{14}$ of the function g at these points.

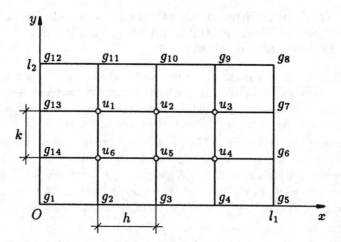

FIGURE 4.1.1

To obtain the unknown values $u_1, \ldots, u_6$ replace, at every inner mesh point, the given Poisson equation by the corresponding difference equation. If we use the formula (1.7.2), we get, at the first point, for example, instead of the derivatives

$$\frac{\partial^2 u}{\partial x^2}$$

$$\frac{\partial^2 u}{\partial y^2}$$

the approximations

$$\frac{\partial^2 u}{\partial x^2} \approx \frac{u_2 - 2u_1 + g_{13}}{h^2}$$

$$\frac{\partial^2 u}{\partial y^2} \approx \frac{g_{11} - 2u_1 + u_6}{k^2}$$

Using similar approximations at other points, we obtain, successively, the following equations at inner mesh points:

$$-\frac{u_2 - 2u_1 + g_{13}}{h^2} - \frac{g_{11} - 2u_1 + u_6}{k^2} = f_1 \qquad (4.1.4)$$

$$-\frac{u_3 - 2u_2 + u_1}{h^2} - \frac{g_{10} - 2u_2 + u_5}{k^2} = f_2$$

$$\cdots\cdots\cdots$$

$$-\frac{u_5 - 2u_6 + g_{14}}{h^2} - \frac{u_1 - 2u_6 + g_2}{k^2} = f_6$$

where $f_1, \ldots, f_6$ are values of the function f at the considered points. In this way we obtain a system of six linear equations for six unknowns $u_1, \ldots, u_6$. This system can be shown to be uniquely solvable. In this way we obtain the wanted finite-difference approximations at inner points of the net. ☐

REMARK 4.1.1

In special cases it is possible to use special procedures. If, for example, $f \equiv 0$ in the problem (4.1.1) and (4.1.2), so that the Poisson equation turns into the Laplace one, and if, moreover, we have a square net (i.e., if $k = h$), then the first equation in the system (4.1.4) becomes

$$-(u_2 - 2u_1 + g_{13}) - (g_{11} - 2u_1 + u_6) = 0$$

that is

$$u_1 = \frac{u_2 + g_{11} + g_{13} + u_6}{4}$$

Similarly, the second (4.1.4) yields

$$u_2 = \frac{u_3 + g_{10} + u_1 + u_5}{4}$$

and so forth. Thus, if the given equation is the Laplace one and if we have a square net, then the value of the net solution at each inner mesh point is equal to the arithmetic mean of values at neighboring points of the net. This fact can be utilized when using the so-called *Liebmann iteration*: We choose arbitrary values $u_1^{(0)}, \ldots, u_6^{(0)}$ as the so-called zero approximation. Then we compute the value $u_1^{(1)}$, the so-called first approximation at point 1, as the arithmetic mean of values at the neighboring points of the net. The value $u_2^{(1)}$ of the first approximation at point 2 is computed similarly, but when computing it, we take into account the "already improved" value $u_1^{(1)}$. In this way we come up to point 6. If necessary (it depends on the accuracy required), we repeat the process to obtain the second approximation $u_1^{(2)}, \ldots, u_6^{(2)}$, and so forth. It can be shown that this method (the Gauss–Seidel method, in essential) converges very rapidly to the exact net solution—the more rapidly, the better the zero approximation has been chosen.

Example 4.1.2
By using the Liebmann iteration, solve the problem

$$-\Delta u = 0 \text{ in } \Omega \tag{4.1.5}$$

$$u = xy \text{ on } \Gamma \tag{4.1.6}$$

where Ω is the square $(0, 3) \times (0, 3)$.

Let us choose $h = 1$ and $k = 1$ (Figure 4.1.2). Taking into account the values of the given function g, indicated at boundary points of the net (see Figure 4.1.2), we choose, at inner mesh points, a "reasonable" zero approximation

$$u_1^{(0)} = 2, \quad u_2^{(0)} = 5, \quad u_3^{(0)} = 2, \quad u_4^{(0)} = 1 \tag{4.1.7}$$

Write these values close to every inner mesh point upward right (as shown in Figure 4.1.2) and make three iteration cycles, writing again values of the first, second, or third approximation close to these points, successively downward right, downward left, or upward left, respectively. We obtain

$$u_1^{(1)} = \frac{5 + 3 + 0 + 1}{4} = 2.25$$

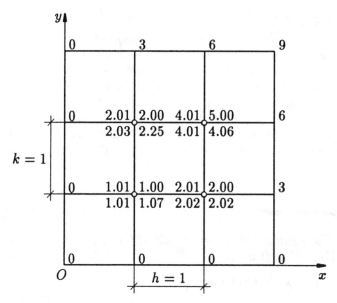

FIGURE 4.1.2

$$u_2^{(1)} = \frac{6 + 6 + 2.25 + 2}{4} \doteq 4.06$$

. . . .

$$u_1^{(2)} = \frac{4.06 + 3 + 0 + 1.07}{4} \doteq 2.03$$

and so forth.

One can see that although we have taken only two decimals into consideration, the iterations converge very rapidly to the values

$$u_1 = 2, \quad u_2 = 4, \quad u_3 = 2, \quad u_4 = 1 \qquad (4.1.8)$$

of the exact net solution. (The solution of the problem (4.1.5) and (4.1.6) is, namely, known, $u = xy$, and its values at inner mesh points are the same as the values (4.1.8) of the net solution. We have chosen a problem with a known solution for the possibility of comparing results.) One can see that the influence of the "wrongly" chosen zero approximation $u_2^{(0)} = 5$ has been quickly eliminated.   ▯

REMARK 4.1.2

The finite-difference method has been explained here for a very simple case only—the equation was the Poisson one and the region was a rectangle. We do not go into detail here, because for elliptical equations this method was overtaken by variational methods in Chapter 3, especially by the finite-element method. The reader who is interested in how to choose difference schemes for equations of higher orders, how to modify the method for regions with curved boundaries, what can be said about convergence of the method, and so forth, finds the needed information, for example, in K. Rektorys 1994, [1] Chapter 27. For problems see Section 4.4.

4.2 The Finite-Difference Method for the Heat Equation

4.2.1 The Explicit Scheme

Let us consider the problem (2.3.1) to (2.3.4), that is, the problem

$$\frac{\partial u}{\partial t} = a^2 \frac{\partial^2 u}{\partial x^2} \quad \text{in } \Omega = (0, l) \times (0, T) \tag{4.2.1}$$

$$u(x, 0) = g(x), \quad 0 < x < l \tag{4.2.2}$$

$$u(0, t) = h_1(t), \quad 0 < t < T \tag{4.2.3}$$

$$u(l, t) = h_2(t), \quad 0 < t < T \tag{4.2.4}$$

(one-dimensional heat conduction without inner heat sources).

On the region Ω, let us construct a net, dividing the interval $[0, l]$ into i subintervals of the length $h = l/i$ and the interval $[0, T]$ into j subintervals of the length $k = T/j$. Denote by $u_1, u_2, u_3,$ and u_4 values of the net solution at the points 1, 2, 3, and 4, as shown in Figure 4.2.1. At point 2 replace Equation (4.2.1) by the corresponding difference equation

$$\frac{u_4 - u_2}{k} = a^2 \frac{u_3 - 2u_2 + u_1}{h^2} \tag{4.2.5}$$

from which

$$u_4 = u_2 + \frac{a^2 k}{h^2} (u_3 - 2u_2 + u_1) \tag{4.2.6}$$

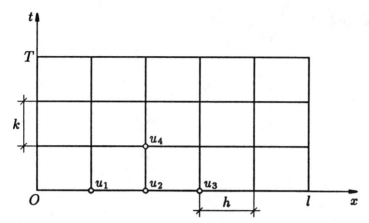

FIGURE 4.2.1

However, a^2, h, and k are known numbers; the values u_1, u_2, and u_3 are known as well, because they are the values of the initial function g at points 1, 2, and 3. Thus, the difference scheme (4.2.5) immediately gives the wanted value u_4. Therefore, this scheme is called *explicit*, in contrast to the one we meet later.

The values in the "first row" of the net (for $t = 0$) are given by the function g. Using (4.2.6), we determine the values of the net solution at every inner mesh point of the second row (for $t = k$). At boundary points of that row they are given by the values of the functions h_1 and h_2. In this way, we obtain values of the net solution in the whole second row, and similarly we can go on: We compute values of the net solution at inner mesh points of the third row (for $t = 2k$); at boundary points of this row these values are given by the functions h_1, h_2, and so forth.

If the condition

$$k < \frac{h^2}{2a^2} \tag{4.2.7}$$

is fulfilled, it can be shown that numerical stability of the process is ensured as well as convergence of the net method for $h \to 0$. (We are not going to say here what we mean by this exactly; see, however, Example 4.1.1.) Moreover, an analogue of the maximum principle (see Theorem 2.3.1) is valid: If on the parabolic part of the boundary (Section 2.3) the values of the net solution lie between the numbers m and M, then the same holds at all mesh points of the region Ω.

Neglecting the condition (4.2.7) can lead to surprising results.

Example 4.2.1

Let us solve the problem

$$\frac{\partial u}{\partial t} = \frac{\partial^2 u}{\partial x^2} \quad \text{in } \Omega = (0, 1) \times (0, 1) \tag{4.2.8}$$

$$u(x, 0) = \sin 5\pi x, \quad 0 < x < 1 \tag{4.2.9}$$

$$u(0, t) = 0, \quad 0 < t < 1 \tag{4.2.10}$$

$$u(1, t) = 0, \quad 0 < t < 1 \tag{4.2.11}$$

The considered problem is a special case of the problem (4.2.1) to (4.2.4) for $l = 1$, $T = 1$, $a^2 = 1$, $g(x) = \sin 5\pi x$, $h_1(t) = 0$, and $h_2(t) = 0$.

Let us choose $h = 0.1$, and $k = 0.1$; at mesh points of the parabolic boundary write the values of the functions g, h_1, and h_2 (see Figure 4.2.2) and determine, using the scheme (4.2.5), that is, formula (4.2.6), values of the net solution at inner mesh points of the second row. If we replace, first, the given equation by difference

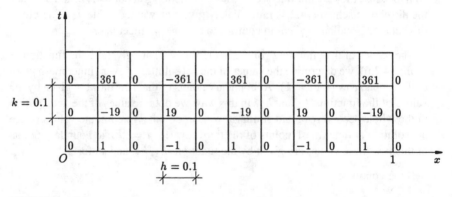

FIGURE 4.2.2

equation (4.2.5) at point (0.1, 0), we obtain by (4.2.6)

$$u_4 = 1 + \frac{0.1}{0.1^2}(0 - 2 + 0) = 1 + 10 \cdot (-2) = -19$$

Similarly, at point (0.2, 0) or (0.3, 0), we obtain (denoting the value of u at the corresponding point (0.2, 0.1) or (0.3, 0.1) by $\tilde{u}_4$ or $\tilde{\tilde{u}}_4$, respectively)

$$\tilde{u}_4 = 0 + \frac{0.1}{0.1^2}(-1 + 2 \cdot 0 + 1) = 0$$

$$\tilde{u}_4 = -1 + \frac{0.1}{0.1^2}(0 + 2 + 0) = -1 + 10 \cdot 2 = 19$$

and so forth. Values of the net solution at all points of the second row are indicated in Figure 4.2.2. At the first inner mesh point of the third row (denote the corresponding value by u_4^*) we then obtain

$$u_4^* = -19 + \frac{0.1}{0.1^2}(0 + 2 \cdot 19 + 0) = -19 + 10 \cdot 38 = 361$$

and so forth (see Figure 4.2.2).   □

Results obtained by the finite-difference method in this way obviously make no sense: The values of the functions g, h_1, and h_2 on the parabolic part of the boundary lie between numbers -1 and $+1$, so that, by Theorem 2.3.1, the values of the solution should lie between these numbers all over the region Ω. The explicit scheme (4.2.5) gives here the values that are essentially larger in absolute value. This result is caused by not respecting the condition (4.2.7) by which, $h = 0.1$ having been chosen, k should satisfy

$$k < \frac{h^2}{2a^2} = \frac{0.1^2}{2} = 0.005$$

This is in contradiction with our choice $k = 0.1$. The reader is advised to carry out the computation for $k = 0.002$, for example, and to compare numerical results with the exact solution

$$u = e^{-25\pi^2 t} \sin 5\pi x$$

of the problem (4.2.8) to (4.2.11).

4.2.2 The Implicit Scheme

The advantage of the explicit scheme consists in applying the formula (4.2.6) that gives explicitly the values of the net solution at inner points of the subsequent row. However, condition (4.2.7) requires the time step of the net to be sufficiently small, which makes the numerical process slow, of course. Therefore, the so-called *implicit scheme* is often used: If we want to get the value u_2 of the net solution at point 2, for example (Figure 4.2.3), we replace the differential equation by the corresponding difference equation at that point, not at the point of the preceding row, as before. In the notation of Figure 4.2.3 we then have

$$\frac{u_2 - u_0}{k} = a^2 \frac{u_3 - 2u_2 + u_1}{h^2} \tag{4.2.12}$$

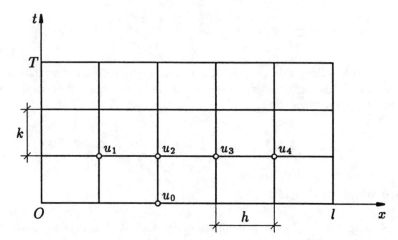

FIGURE 4.2.3

Here, simultaneously with u_2, also the values u_1 and u_3 are unknown. To be able to determine the net solution at all points 1, 2, 3, and 4 of the considered row, we have to write difference equations, similar to Equation (4.2.12), also at points 1, 3, and 4. In this way (values at the points $(0, k)$ and (l, k) being known), we obtain a system of four (linear) equations for four unknowns $u_1, \ldots, u_4$. This system can be shown to be uniquely solvable. Similarly we proceed further. All essential properties concerning stability and convergence of the process remain preserved here without regard to whether the condition (4.2.7) is fulfilled. Thus the time step k may be chosen arbitrarily. In return, it is necessary to solve a system of equations for every row.

In both cases, of the explicit as well as of the implicit scheme, computing programs can be easily prepared. As a rule, the implicit scheme is more advantageous. See also Section 4.4.

REMARK 4.2.1   *(the finite-difference method for the wave equation)*
   Let us solve the problem

$$\frac{\partial^2 u}{\partial t^2} = a^2 \frac{\partial^2 u}{\partial x^2} \quad \text{in } \Omega = (0, l) \times (0, T) \tag{4.2.13}$$

$$u(x, 0) = g_1(x), \quad 0 < x < l \tag{4.2.14}$$

$$\frac{\partial u}{\partial t}(x, 0) = g_2(x), \quad 0 < x < l \tag{4.2.15}$$

$$u(0, t) = u(l, t) = 0 , \quad 0 < t < T \tag{4.2.16}$$

(Vibration of a perfectly flexible fiber (cf. Equation (2.1.17) and following) has an initial amplitude $g_1(x)$, an initial velocity $g_2(x)$, and its ends are fixed.) Here, the following explicit scheme can be used (for the notation see Figure 4.2.4)

$$u_4 = u_2 + u_3 - u_1 \tag{4.2.17}$$

with

$$k = \frac{h}{a} \tag{4.2.18}$$

The formula (4.2.17) requires values of the net solution in the preceding two

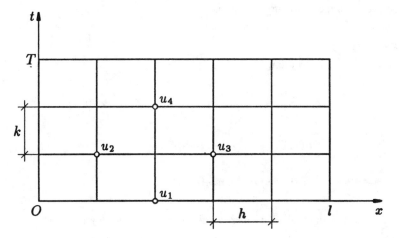

FIGURE 4.2.4

rows to be known. In the first row (for $t = 0$), these values are given by the function $g_1(x)$. At inner points of the second row (for $t = k = h/a$) we obtain them as values of the function

$$g_1 + kg_2 = g_1 + \frac{h}{a} g_2 \tag{4.2.19}$$

(using the fact that for $t = 0$ the values of the function u as well as of its derivative with respect to t are known). By (4.2.17) we then compute values of the net solution at inner points of the third row, and so forth.

For details and for schemes for the wave equation in two variables see, for example, K. Rektorys 1994, [1] Chapter 27. See also Problem 4.4.8.

4.3 The Method of Discretization in Time (the Rothe Method, the Method of Lines)

The characteristic feature of the finite-difference method is that the given equation is discretized with respect to all variables, so that all derivatives in that equation are replaced by the corresponding difference quotients.

Often methods in numerical mathematics are used, where only one variable, or some of them, are discretized, which makes it possible not only to solve effectively the given problems, but also to get, at the same time, a better insight into the structure of their solutions. We are going to show the case here where the time t is discretized, thus, the case of the so-called *method of discretization in time* (the *Rothe method*, also the *[horizontal] method of lines*). The idea of this method will be shown in the following example.

Example 4.3.1

Let us solve the problem

$$\frac{\partial u}{\partial t} = \frac{\partial^2 u}{\partial x^2} + \sin x \quad \text{in} \quad \Omega = (0, \pi) \times (0, 1) \tag{4.3.1}$$

$$u(x, 0) = 0, \quad 0 < x < \pi \tag{4.3.2}$$

$$u(0, t) = u(\pi, t) = 0, \quad 0 < t < 1 \tag{4.3.3}$$

Divide the interval $[0, 1]$ into p subintervals of the length $h = 1/p$ (in Figure 4.3.1 $p = 3$ has been chosen) and find, successively, for $t_1 = h, t_2 = 2h, \ldots, t_p = ph$ the functions

$$u_1(x), u_2(x), \ldots, u_p(x) \tag{4.3.4}$$

so that the relations

$$\frac{u_1 - u_0}{h} = u_1'' + \sin x \tag{4.3.5}$$

$$u_1(0) = 0, \quad u_1(\pi) = 0 \tag{4.3.6}$$

$$\frac{u_2 - u_1}{h} = u_2'' + \sin x \tag{4.3.7}$$

$$u_2(0) = 0, \quad u_2(\pi) = 0 \tag{4.3.8}$$

. . . .

$$\frac{u_p - u_{p-1}}{h} = u_p'' + \sin x \tag{4.3.9}$$

$$u_p(0) = 0 , \quad u_p(\pi) = 0 \tag{4.3.10}$$

are satisfied, while $u_0 \equiv 0$ by (4.3.2). The derivative $\partial u/\partial t$ is thus replaced, at

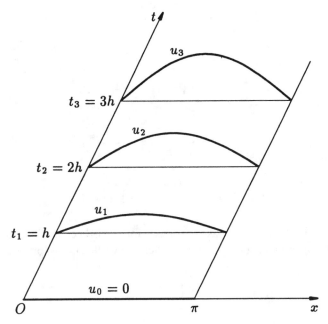

FIGURE 4.3.1

every point t_j, $j = 1, 2, \ldots, p$, by the difference quotient

$$\frac{u_j - u_{j-1}}{h}$$

and the problem (4.3.1) to (4.3.3) is converted, in this way, into a finite sequence of problems (4.3.5), (4.3.6), ..., (4.3.10) in ordinary differential equations, to be solved successively.

Let us write the problem (4.3.5) and (4.3.6) in the form (recall that $u_0 = 0$)

$$-u_1'' + \frac{u_1}{h} = \sin x \tag{4.3.11}$$

$$u_1(0) = 0, \quad u_1(\pi) = 0 \qquad (4.3.12)$$

This problem is easily solved: In fact, u_1 can be obviously assumed in the form

$$u_1 = c_1 \sin x \qquad (4.3.13)$$

Putting this into Equation (4.3.11), we obtain

$$c_1 \sin x + \frac{c_1 \sin x}{h} = \sin x$$

from which

$$c_1 \left(1 + \frac{1}{h} \right) = 1$$

and

$$c_1 = \frac{h}{1+h} = 1 - \frac{1}{1+h}$$

so that

$$u_1 = \left(1 - \frac{1}{1+h} \right) \sin x \qquad (4.3.14)$$

(Why the form $1 - 1/(1+h)$ has been chosen for c_1 will become clear immediately.)
Similarly, the problem (4.3.7) and (4.3.8) can be written in the form (using 4.3.14)

$$-u_2'' + \frac{u_2}{h} = \frac{\left(1 - \frac{1}{1+h} \right) \sin x}{h} + \sin x$$

$$u_2(0) = 0$$

$$u_2(\pi) = 0$$

Assuming u_2 to be of the form

$$u_2 = c_2 \sin x$$

we get, as before,

$$c_2 \sin x + \frac{c_2 \sin x}{h} = \frac{(1+h) \sin x - \frac{1}{1+h} \sin x}{h}$$

which yields

$$\frac{1+h}{h} c_2 = \frac{1+h}{h} - \frac{1}{h(1+h)}$$

so that

$$c_2 = 1 - \frac{1}{(1+h)^2}$$

and

$$u_2 = \left(1 - \frac{1}{(1+h)^2}\right) \sin x$$

In general, we get

$$u_j = \left(1 - \frac{1}{(1+h)^j}\right) \sin x$$

$$j = 1, 2, \ldots, p$$

If we choose, in particular, $p = 5$, so that $h = 0.2$, by rounding off to two decimals we obtain

$$u_1 = 0.17 \sin x \qquad (4.3.15)$$

$$u_2 = 0.31 \sin x$$

$$u_3 = 0.42 \sin x$$

$$u_4 = 0.52 \sin x$$

$$u_5 = 0.60 \sin x$$

The exact solution of the given problem is known

$$u = \left(1 - e^{-t}\right) \sin x \qquad (4.3.16)$$

which enables us to compare the approximate solution with the exact one: For $t = 0.2, t = 0.4, t = 0.6, t = 0.8$, and $t = 1$ (4.3.16) yields

$$u(x, \ 0.2) = 0.18 \sin x \ , \qquad u(x, \ 0.4) = 0.33 \sin x \qquad (4.3.17)$$

$$u(x, \ 0.6) = 0.45 \sin x \ , \qquad u(x, \ 0.8) = 0.55 \sin x \ , \qquad u(x, \ 1) = 0.63 \sin x$$

Although step $h = 0.2$ of the division is rather rough, the approximate solution (4.3.15) is in a good accordance with the exact one. In K. Rektorys 1982, [3] Chapter 1, it is shown that choosing $h = 0.01$, the result obtained by the method of discretization in time at the points $t_1, t_2, \ldots, t_{100}$ is almost identical with the exact solution.   ☐

REMARK 4.3.1

The problem (4.3.1) to (4.3.3) is so simple that it is not necessary to apply to its solution any approximate method, for example, just the method of discretization in time. The advantage of that method becomes clear even when solving much more complicated problems of the form

$$\frac{\partial u}{\partial t} + Au = f$$

where A is an elliptical operator of the type considered in Chapter 3 (thus, when solving, for example, the equation

$$\frac{\partial u}{\partial t} - \Delta u = f$$

for two-dimensional heat conduction on a not very simple region, and with, perhaps not very simple boundary conditions). The method of discretization in time leads then to successive solution of elliptical problems, where variational methods, developed in Chapter 3, can be applied with success. The method of discretization in time also can be applied to the solution of hyperbolic problems (e.g., occurring in dynamics). For details, see K. Rektorys 1982, [3] where this method is discussed from the theoretical as well as from the numerical point of view (existence theorems, error estimates, etc.), and where many numerical examples also are given. See also Section 4.4.

Instead of discretizing time, when solving partial differential equations, the space variable (or space variables) can be discretized. In this connection one often speaks about the (vertical) method of lines, or about the *Galerkin method* (in a sense different from that mentioned in Section 3.5). This method then leads to the solution of a system of ordinary differential equations for functions depending on time only.

4.4   Problems 4.4.1 to 4.4.9

4.4.1—(mixed boundary conditions in the Poisson problem). By using the finite-difference method with step $k = \frac{1}{3}$ in the direction of the x- and y-axis solve the problem

$$-\Delta u = (\pi^2 y^2 - 2)\sin \pi x \quad \text{in} \quad \Omega = (0, 1) \times (0, 1) \qquad (4.4.1)$$

$$u = 0 \quad \text{for} \quad x = 0, \quad 0 \leqq y \leqq 1 \qquad (4.4.2)$$

$$u = 0 \quad \text{for} \quad x = 1, \quad 0 \leqq y \leqq 1 \qquad (4.4.3)$$

$$u = \sin \pi x \quad \text{for} \quad y = 1, \quad 0 < x < 1 \qquad (4.4.4)$$

$$\frac{\partial u}{\partial v} = -\frac{\partial u}{\partial y} = 0 \quad \text{for} \quad y = 0, \quad 0 < x < 1 \qquad (4.4.5)$$

Hint: (See Figure 4.4.1.) To characterize condition (4.4.5), choose, at the mesh points of the x-axis, the same values for the approximate solution as on line $y = \frac{1}{3}$ (see the figure). For example, at point $(\frac{1}{3}, \frac{1}{3})$, you then get

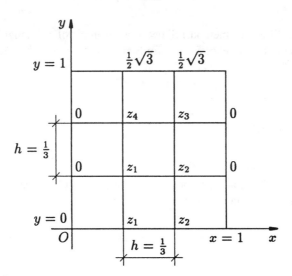

FIGURE 4.4.1

$$-\frac{z_2 - 2z_1 + 0}{\left(\frac{1}{3}\right)^2} - \frac{z_4 - 2z_1 + z_1}{\left(\frac{1}{3}\right)^2} = \left(\frac{\pi^2}{9} - 2\right)\frac{\sqrt{3}}{2}$$

In this way you obtain a system of four equations for the unknowns $z_1, \ldots, z_4$. The result is

$$z_1 = z_2 \doteq 0.167 , \quad z_3 = z_4 \doteq 0.421 \tag{4.4.6}$$

4.4.2—The explicit solution of the problem (4.4.1) to (4.4.5) is known

$$u = y^2 \sin \pi x$$

Compare the values of this solution (at corresponding points) with approximations (4.4.6).

[We have

$$u_1 = u_2 \doteq 0.096 \tag{4.4.7}$$

$$u_3 = u_4 \doteq 0.385$$

The difference is considerable, especially as concerns the values u_1, u_2 and z_1, z_2. Something like this could be expected. Why?]

4.4.3—(Boundary conditions for the equation

$$\Delta^2 u = f \tag{4.4.8}$$

in the finite-difference method.) Show on an example of a rectangular region how to choose a finite-difference scheme for Equation (4.4.8) with boundary conditions

$$u = 0 \tag{4.4.9}$$

$$\frac{\partial u}{\partial v} = 0$$

(deflection of a clamped plate).

[See Figure 4.4.2; at inner mesh points, the values of the net solution can then be completed following the scheme shown in Figure 4.4.3.

$$u_0 = \frac{1}{20}[8(u_1 + u_2 + u_3 + u_4) - 2(u_5 + u_6 + u_7 + u_8)$$

$$- (u_9 + u_{10} + u_{11} + u_{12})] + \frac{1}{20} f_0 h^4 \bigg]$$

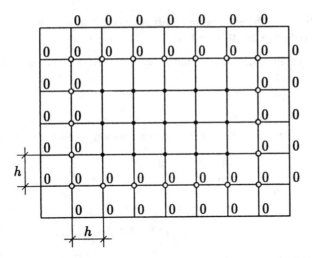

FIGURE 4.4.2

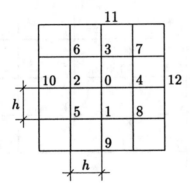

FIGURE 4.4.3

4.4.4—Solve the problem

$$\frac{\partial u}{\partial t} = \frac{\partial^2 u}{\partial x^2} \quad \text{in } \Omega \quad (0 < x < 1, \quad t > 0) \tag{4.4.10}$$

$$u(x, 0) = \sin 2\pi x \tag{4.4.11}$$

$$u(0, t) = 0 \tag{4.4.12}$$

$$u(1, t) = 0$$

1. By using the explicit finite-difference scheme with $h = 0.25$, $k = 0.01$, verify whether the condition (4.2.7) is satisfied

2. By the implicit scheme with $h = 0.25$, $k = 0.03$, compare the results for $t = 0.03$

Hint:

1. Conditions (4.2.7) satisfied; in the notation of Figure 4.2.1 we have, for $x = 0.25$

$$\frac{u_4 - u_2}{0.01} = \frac{1}{\left(\frac{1}{4}\right)^2} (u_3 - 2u_2 + u_1)$$

that is

$$u_4 = 1 + 0.16(0 - 2 + 0) = 0.680$$

and so forth (see Figure 4.4.4).

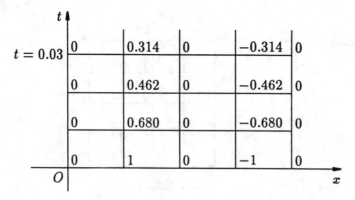

FIGURE 4.4.4

2. Here we have, in the notation of Figure 4.2.3,

$$\frac{u_1 - 1}{0.03} = \frac{1}{\left(\frac{1}{4}\right)^2} (u_2 - 2u_1 + 0)$$

that is,

$$1.96u_1 - 0.48u_2 + 0 \cdot u_3 = 1$$

and so forth. For numerical values see Figure 4.4.5. The difference between (1) and (2) is considerable. See also Problems 4.4.5 and 4.4.6.

4.4.5—Solve the problem (4.4.10) to (4.4.12) by the Rothe method

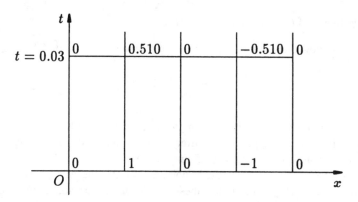

FIGURE 4.4.5

1. With the step $k = 0.01$
2. With the step $k = 0.03$

Hint:
1.

$$\frac{u_1(x) - u(x, 0)}{0.01} = u_1''$$

Assuming the solution in the form

$$u_1(x) = A_1 \sin 2\pi x$$

we obtain

$$u_1(x) = 0.717 \sin 2\pi x$$

Further

$$u_2(x) = 0.514 \sin 2\pi x$$

$$u_3(x) = 0.368 \sin 2\pi x$$

2.

$$\frac{v_1(x) - u(x, 0)}{0.03} = v_1''$$

$$v_1(x) = 0.458 \sin 2\pi x$$

(cf. Problem 4.4.6.)

4.4.6—Show that explicit solution of the problem (4.4.10) to (4.4.12) is

$$u = e^{-4\pi^2 t} \sin 2\pi x \qquad (4.4.13)$$

This result makes it possible to compare efficiency of the Rothe method and the finite-difference one, applied with different steps. Evidently it is sufficient to compare numerical values for $x_1 = 0.25$ only. For $t_1 = 0.01$, $t_2 = 0.02$, or $t_3 = 0.03$, the exact solution (4.4.13) gives, at point $x_1 = 0.25$, the value $0.674, 0.454$, or 0.306, respectively. Comparing these results with corresponding results (for $x = x_1$) in Problems 4.4.4 and 4.4.5, we see that the best approximation has been obtained by the explicit finite-difference method.

4.4.7—(Discontinuity in boundary conditions, see (2.3.5).) Solve the problem

$$\frac{\partial u}{\partial t} = \frac{\partial^2 u}{\partial x^2} \quad \text{in } \Omega \quad (0 < x < 1, \quad t > 0)$$

$$u(x, 0) = 1$$

$$u(0, t) = 0$$

$$u(1, t) = 0$$

1. By the explicit finite-difference method with $h = 0.25$ and $k = 0.01$

2. By the Rothe method with the same time step, combined with the Ritz method with $v_1 = x - x^2$

In both cases (1) and (2) give numerical results for three steps, successively.

[1. See Figure 4.4.6

2. See Figure 4.4.7]

4.4.8—By using the finite-difference scheme (4.2.17) and (Figure 4.2.4), solve the problem

$$\frac{\partial^2 u}{\partial t^2} = 4\frac{\partial^2 u}{\partial x^2} \quad \text{in } \Omega \quad (0 < x < 1, \quad t > 0)$$

$$u(x, 0) = x - x^2$$

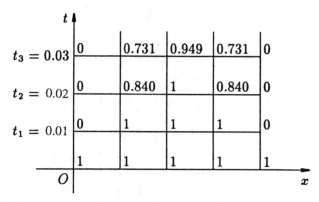

FIGURE 4.4.6

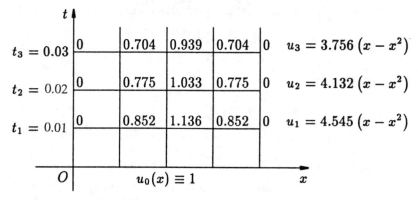

FIGURE 4.4.7

$$\frac{\partial u}{\partial t}(x, 0) = \sin \pi x$$

$$u(0, t) = 0$$

$$u(1, t) = 0$$

Hint:

See Figure 4.4.8; for $t = 0$, numerical values of the initial function $x - x^2$ are given at points $x_1 = \frac{1}{4}$, $x_2 = \frac{2}{4}$, and $x_3 = \frac{3}{4}$; by (4.2.18) we have $k = \frac{1}{2}h = \frac{1}{8}$; for $t = \frac{1}{8}$, the values 0.276, and so forth have been calculated using (4.2.19). Pay attention to periodicity (in time) of the solution; cf. also Problem 5.2.4.

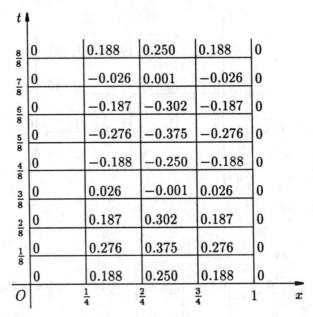

FIGURE 4.4.8

4.4.9*—(The Rothe method applied to the solution of a more-dimensional problem) By the Rothe method with the step $k = 0.01$ solve the problem

$$\frac{\partial u}{\partial t} - \left[\frac{\partial}{\partial x}\left(\left(4 + \frac{y^2}{8}\right)\frac{\partial u}{\partial x}\right) + \frac{\partial}{\partial y}\left(\left(4 + \frac{x^2}{25}\right)\frac{\partial u}{\partial y}\right)\right] \quad (4.4.14)$$

$$= 1 - \frac{x^2}{16} - \frac{y^4}{4} \text{ in } \Omega \times (0, 1)$$

$$u(x, y, 0) = 0 \quad (4.4.15)$$

$$u = 0 \text{ on } \Gamma \times (0, 1) \quad (4.4.16)$$

where Ω is the interior of the ellipse Γ with half-axis $a = 4$, or $b = 2$ in the positive x- or y-axis, respectively. Carry out two time steps, assuming the Ritz approximations u_1^*, or u_2^*, of the Rothe functions u_1, or u_2, to be of the form

$$u_1^* = a_1 \underbrace{\left(1 - \frac{x^2}{16} - \frac{y^2}{4}\right)}_{v_1}$$

$$u_2^* = a_2 \underbrace{\left(1 - \frac{x^2}{16} - \frac{y^2}{4}\right)}_{v_1}$$

respectively.

Hint:

Verify, first, that operator A given by

$$Au = -\frac{\partial}{\partial x}\left(\left(4 + \frac{y^2}{8}\right)\frac{\partial u}{\partial x}\right) - \frac{\partial}{\partial y}\left(\left(4 + \frac{x^2}{25}\right)\frac{\partial u}{\partial y}\right)$$

is positive (and thus symmetrical) on its domain of definition D_A. At the same time, the Green theorem gives

$$(Au, v) = \int\int_\Omega \left[\left(4 + \frac{y^2}{8}\right)\frac{\partial u}{\partial x}\frac{\partial v}{\partial x} + \left(4 + \frac{x^2}{25}\right)\frac{\partial u}{\partial y}\frac{\partial v}{\partial y}\right] dxdy$$

Applying the Rothe method, we obtain, first

$$\frac{u_1(x, y) - 0}{0.01} + Au_1 = 1 - \frac{x^2}{16} - \frac{y^2}{4}$$

Operator A being positive, as is operator B, given by

$$Bu = Au + \frac{u}{0.01}$$

The Ritz method with

$$u_1^* = a_1 v_1 = a_1\left(1 - \frac{x^2}{16} - \frac{y^2}{4}\right)$$

then yields

$$(Bv_1, v_1)\, a_1 = (f, v_1) \tag{4.4.17}$$

Using semipolar coordinates

$$x = 4r \cos \phi \qquad (4.4.18)$$

$$y = 2r \sin \phi$$

(see, e.g., K. Rektorys 1994, [1] Section 14.4), we obtain

$$(Bv_1, v_1) = \int\int_\Omega \left[\left(4 + \frac{y^2}{8} \right) \left(\frac{\partial v_1}{\partial x} \right)^2 \right.$$

$$+ \left(4 + \frac{x^2}{25} \right) \left(\frac{\partial v_1}{\partial y} \right)^2 + \frac{1}{0.01} v_1^2 \Bigg] dxdy$$

$$= \int\int_\Omega \left[\left(4 + \frac{y^2}{8} \right) \frac{x^2}{64} + \left(4 + \frac{x^2}{25} \right) \frac{y^2}{4} \right.$$

$$+ \frac{1}{0.01} \left(1 - \frac{x^2}{16} - \frac{y^2}{4} \right)^2 \Bigg] dxdy$$

$$= 4 \cdot 2 \int_0^1 \int_0^{2\pi} \left[\left(4 + \frac{4r^2 \sin^2 \phi}{8} \right) \cdot \frac{16r^2 \cos^2 \phi}{64} \right.$$

$$+ \left(4 + \frac{16r^2 \cos^2 \phi}{25} \right) \cdot \frac{4r^2 \sin^2 \phi}{4} \Bigg] rdrd\phi$$

$$+ \frac{1}{0.01} \cdot 4 \cdot 2 \int_0^1 \int_0^{2\pi} (1 - r^2)^2 rdrd\phi$$

$$= 8 \left(\frac{\pi}{4} + \frac{1}{6 \cdot 8} \cdot \frac{\pi}{4} + \pi + \frac{16}{6 \cdot 25} \cdot \frac{\pi}{4} \right) + \frac{8}{0.01} \left(\frac{1}{2} - \frac{2}{4} + \frac{1}{6} \right) \cdot 2\pi$$

$$\doteq 32.217 + \frac{1}{0.01} \cdot 8.378$$

With the help of (4.4.18) we then easily obtain

$$(f, v_1) = \int \int_{\Omega} \left(1 - \frac{x^2}{16} - \frac{y^2}{4} \right)^2 dxdy \doteq 8.378$$

By (4.4.17) we have

$$a_1 \doteq \frac{8.378}{870} \doteq 0.0096$$

and

$$u_1^* \doteq 0.0096 \left(1 - \frac{x^2}{16} - \frac{y^2}{4} \right)$$

Having obtained u_1^*, we get (in a much easier way, now)

$$u_2^* \doteq 0.0189 \left(1 - \frac{x^2}{16} - \frac{y^2}{4} \right)$$

Chapter 5

The Fourier Method

5.1 The Fourier Method for One-Dimensional Vibration Problems

The *Fourier method* is a very popular method of solving engineering or physical problems, although its application is restricted to the solution of relatively special problems, as one can see in the following example.

Let us consider the problem

$$\frac{\partial^2 u}{\partial t^2} = a^2 \frac{\partial^2 u}{\partial x^2} \quad \text{in} \quad \Omega = (0, l) \times (0, T) \tag{5.1.1}$$

$$u(x, 0) = g(x) \tag{5.1.2}$$

$$0 < x < l$$

$$\frac{\partial u}{\partial t}(x, 0) = 0 \tag{5.1.3}$$

$$0 < x < l$$

$$u(0, t) = u(l, t) = 0 \tag{5.1.4}$$

$$0 < t < T$$

(Cf. Remark (4.2.1); Vibration of a perfectly flexible fiber, with initial amplitude $g(x)$, zero initial velocity and with its ends fixed.)

Let us assume the solution in the form

$$u(x, t) = \sum_{n=1}^{\infty} b_n u_n(x, t) \tag{5.1.5}$$

where each of the functions $u_n(x, t)$

1.   Is of the form $u_n(x, t) = X_n(x)T_n(t)$

2.   Fulfills Equation (5.1.1)

3.   Satisfies conditions (5.1.3) and (5.1.4)

4.   Is not identically zero in the region Ω.

Thus every finite (= partial) sum of the series (5.1.5) will fulfill Equation (5.1.1) and conditions (5.1.3) and (5.1.4). By a proper choice of the coefficients b_n we then fulfill the condition (5.1.2).

   First, we have to determine the form of the functions u_n.

   By (1) we have

$$\frac{\partial^2 u_n}{\partial x^2} = X_n'' T_n \tag{5.1.6}$$

$$\frac{\partial^2 u_n}{\partial t^2} = X_n \ddot{T}_n$$

where by a dash or dot the derivative with respect to x or t is denoted, respectively. By condition (2) we thus have

$$X_n \ddot{T}_n = a^2 X_n'' T_n \text{ in } \Omega \tag{5.1.7}$$

Let us divide Equation (5.1.7) by the product $a^2 X_n T_n$, assuming $a^2 X_n T_n \neq 0$ in Ω. (This assumption will turn out later to be unessential.) We obtain

$$\frac{1}{a^2} \frac{\ddot{T}_n}{T_n} = \frac{X_n''}{X_n} \text{ in } \Omega \tag{5.1.8}$$

This equation should be fulfilled in Ω, that is, for every $x \in (0, l)$ and for every $t \in (0, T)$. However, the left-hand side of this equation does not depend on x (T_n being a function of t only), the right-hand side does not depend on t. Thus, the left-hand as well as the right-hand side of that equation should be equal to a constant. Denote it by $-\lambda_n$, that is

$$\frac{1}{a^2} \frac{\ddot{T}_n}{T_n} = \frac{X_n''}{X_n} = -\lambda_n \qquad (5.1.9)$$

Next consider the second of equalities (5.1.9) and rewrite it in the form

$$X_n'' + \lambda_n X_n = 0 \qquad (5.1.10)$$

This is a differential equation of the second order for the function $X_n(x)$. Condition (3) then yields

$$X_n(0) = 0 \qquad (5.1.11)$$

$$X_n(l) = 0$$

(If, namely, we had $X_n(0) \neq 0$, e.g., then because of condition (5.1.4) the function $T_n(t)$ had to be equal to zero for all t considered, what would yield $u_n(x, t) \equiv 0$ in contradiction to (4).)

Condition (4) further implies that the function $X_n(x)$ must not be identically equal to zero. However, then the problem (5.1.10) and (5.1.11) is the well-known eigenvalue problem (Sections 1.4 and 1.5). Nonzero solutions exist only if

$$\lambda_n = \frac{n^2 \pi^2}{l^2} \qquad (5.1.12)$$

$$n = 1, 2, \ldots$$

and corresponding eigenfunctions are (up to a multiplicative constant)

$$X_n = \sin \frac{n \pi x}{l} \qquad (5.1.13)$$

$$n = 1, 2, \ldots$$

In this way, the form of the functions $X_n(x)$ has been determined. Let us find the corresponding functions $T_n(t)$.

Using (5.1.12), the first equality (5.1.9)

$$\frac{1}{a^2} \frac{\ddot{T}_n}{T_n} = -\lambda_n$$

gives

$$\ddot{T}_n + \frac{a^2 n^2 \pi^2}{l^2} T_n = 0 \tag{5.1.14}$$

The general solution of this equation is

$$T_n = C_1 \cos \frac{a n \pi t}{l} + C_2 \sin \frac{a n \pi t}{l} \tag{5.1.15}$$

Condition (3) (requiring that the function $u_n(x, t)$ satisfies condition (5.1.3) yields

$$\dot{T}_n(0) = 0 \tag{5.1.16}$$

(Otherwise the function $X_n(x)$ should be identically equal to zero; cf. to the above similar consideration.) Now, by (5.1.15) we have

$$\dot{T}_n = -\frac{a n \pi}{l} \sin \frac{a n \pi t}{l} \cdot C_1 + \frac{a n \pi}{l} \cos \frac{a n \pi t}{l} \cdot C_2$$

and thus

$$\dot{T}_n(0) = \frac{a n \pi}{l} C_2$$

The condition (5.1.16) then implies

$$C_2 = 0$$

because $a n \pi / l \neq 0$. If we choose the coefficient C_1 in (5.1.15) equal to 1 (so that the determination of the coefficients b_n later is as simple as possible), we get

$$T_n = \cos \frac{a n \pi t}{l} \tag{5.1.17}$$

and thus

$$u_n(x, t) = \sin \frac{n \pi x}{l} \cos \frac{a n \pi t}{l} \tag{5.1.18}$$

Each of the functions (5.1.18), $n = 1, 2, \ldots$, satisfies (without regard to whether $X_n T_n \neq 0$, see (5.1.7) and (5.1.8) differential Equation (5.1.1) and the conditions (5.1.3) and (5.1.4), and the same is valid for an arbitrary linear combination of them.

Let the solution $u(x, t)$ of the problem (5.1.1) and (5.1.4) have the form (5.1.5). Since for $t = 0$

$$u_n(x, 0) = \sin \frac{n \pi x}{l}$$

it follows from conditions (5.1.2) that

$$g(x) = \sum_{n=1}^{\infty} b_n \sin \frac{n\pi x}{l} \tag{5.1.19}$$

The wanted coefficients b_n are then the Fourier coefficients of the function $g(x)$ with respect to the system of functions $\sin(n\pi x/l)$, and, consequently (K. Rektorys 1994, [1] Section 16.3; cf. also Section 1.3 of this book)

$$b_n = \frac{2}{l} \int_0^l g(x) \sin \frac{n\pi x}{l} \, dx \tag{5.1.20}$$

$$n = 1, 2, \ldots$$

Thus, finally

$$u(x, t) = \sum_{n=1}^{\infty} b_n \sin \frac{n\pi x}{l} \cos \frac{an\pi t}{l} \tag{5.1.21}$$

with b_n given by (5.1.20).

Here, it is necessary to point out that the solution (5.1.21) is a formal one only. If we take a finite number of terms of the series (5.1.21), then this finite sum fulfills Equation (5.1.1) and conditions (5.1.3) and (5.1.4). However, for the sum of the infinite series this assertion does not need to be true; in general (i.e., for a general function g), the sum of this series does not need to have even the second derivatives with respect to x and t, as required by differential Equation (5.1.1). It can be shown that, for the function $u(x, t)$ given by the series (5.1.21) to be a classical solution of the given problem, it is sufficient if the function g and its derivatives up to the second order are continuous in $[0, l]$ and if this function fulfills

$$g(0) = g(l) = 0 \tag{5.1.22}$$

$$g''(0) = g''(l) = 0$$

If these conditions are not fulfilled, (5.1.21) does not need to be a (classical) solution. However, from the engineering or physical point of view, it is of no particular harm. An engineer or physicist takes, in practice, a finite number of terms in (5.1.21) only

$$u_k(x, t) = \sum_{n=1}^{k} b_n \sin \frac{n\pi x}{l} \cos \frac{an\pi t}{l} \tag{5.1.23}$$

This function then satisfies exactly the given differential equation and conditions (5.1.3) and (5.1.4); it does not need to fulfill the condition (5.1.2), because in the series (5.1.19) for the function g only a finite number of terms has been taken. This inaccuracy can be tolerated, in practice, as usual (in (5.1.23) we have to take k sufficiently large).

REMARK 5.1.1

If instead of the condition (5.1.3) the condition

$$\frac{\partial u}{\partial t}(x, 0) = h(x) \tag{5.1.24}$$

$$0 < x < l$$

is given, we obtain, instead of the series (5.1.21), the series

$$\sum_{n=1}^{\infty} \sin \frac{n\pi x}{l} \left(b_n \cos \frac{an\pi t}{l} + c_n \sin \frac{an\pi t}{l} \right) \tag{5.1.25}$$

with coefficients b_n given (5.1.20) and c_n given by

$$c_n = \frac{2}{an\pi} \int_0^l h(x) \sin \frac{n\pi x}{l} \, dx \tag{5.1.26}$$

For the case of nonhomogeneous boundary conditions see Problems 5.2.5 and 5.2.6.

REMARK 5.1.2

Further application of the Fourier method to the solution of a heat-conduction problem in an infinite cylinder, and so forth can be found in K. Rektorys 1994, [1] Chapter 26. See also Section 5.2.

5.2   Problems 5.2.1 to 5.2.8

5.2.1—By using the Fourier method, solve the problem

$$\frac{\partial u}{\partial t} = a^2 \frac{\partial^2 u}{\partial x^2} \quad \text{in } \Omega \quad (0 < x < l, \quad t > 0)$$

$$u(x, 0) = g(x)$$

$$u(0, t) = u(l, t) = 0$$

(cf. also Problem 5.2.2).

$$\left[u(x, t) = \sum_{n=1}^{\infty} a_n \, e^{-\frac{a^2 n^2 \pi^2 t}{l^2}} \sin \frac{n \pi x}{l} \right.$$

where

$$\left. a_n = \frac{2}{l} \int_0^l g(x) \sin \frac{n \pi x}{l} \, dx \right]$$

5.2.2—Solving the problem

$$\frac{\partial u}{\partial t} = \frac{\partial^2 u}{\partial x^2} \quad \text{in} \quad \Omega \quad (0 < x < 1, \quad t > 0)$$

$$u(x, 0) = \sin 2\pi x$$

$$u(0, t) = u(l, t) = 0$$

by the Fourier method, we obtain its exact solution (4.4.13),

$$u = e^{-4\pi^2 t} \sin 2\pi x$$

Why?

Hint: The system of functions $\sin n\pi x$, $n = 1, 2, \ldots$ is orthogonal in $L_2(0, 1)$.

5.2.3—Solve the problem

$$\frac{\partial u}{\partial t} = \frac{\partial^2 u}{\partial x^2} \quad \text{in} \quad \Omega \quad (0 < x < 1, \quad t > 0)$$

$$u(x, 0) \equiv 1$$

$$u(0, t) = u(1, t) = 0$$

(for discontinuous boundary conditions, see Problem 4.4.7) by the Fourier method. Compare the "exact" solution obtained with approximate solutions obtained in

Problem 4.4.7 (1) and (2) by the finite-difference and the Rothe methods, taking three terms of the series into account.

[We have

$$u(x, t) = \frac{4}{\pi} + \tag{5.2.1}$$

$$\left(e^{-\pi^2 t} \sin \pi x + \frac{e^{-9\pi^2 t} \sin 3\pi x}{3} + \frac{e^{-25\pi^2 t} \sin 5\pi x}{5} + \cdots \right)$$

Comparing numerical results obtained in our problem with those obtained in Problem 4.4.7 (1) and (2) and taking into account that they have been obtained by quite different methods and influenced, moreover, by discontinuities in boundary conditions, we can conclude that they correspond relatively well. The main differences are at the points of discontinuities $(0, 0)$ and $(0, 1)$ of boundary conditions (see Figure 5.2.1).]

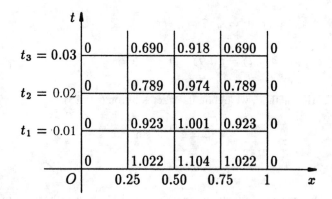

FIGURE 5.2.1

5.2.4—Solve the problem

$$\frac{\partial^2 u}{\partial t^2} = 4 \frac{\partial^2 u}{\partial x^2} \quad \text{in } \Omega \quad (0 < x < 1, \quad t > 0)$$

$$u(x, 0) = x - x^2$$

$$\frac{\partial u}{\partial t}(x, 0) = \sin \pi x$$

$$u(0, t) = u(1, t) = 0$$

by the Fourier method and compare the results for points $x_1 = \frac{1}{4}$, $x_2 = \frac{2}{4}$, $x_3 = \frac{3}{4}$; and points $t_0 = 0$, $t_1 = \frac{1}{8}$, $t_2 = \frac{2}{8}$, $t_3 = \frac{3}{8}$, and $t_4 = \frac{4}{8}$ with those obtained in Problem 4.4.8.

Hint: We have

$$x - x^2 = \frac{8}{\pi^3} \left(\frac{\sin \pi x}{1^3} + \frac{\sin 3\pi x}{3^3} + \frac{\sin 5\pi x}{5^3} + \frac{\sin 7\pi x}{7^3} + \cdots \right)$$

(cf. K. Rektorys 1994, [1] Section 16.3); by (5.1.25) we then get

$$u = \frac{8}{\pi^3} \left(\frac{\sin \pi x \cos 2\pi t}{1^3} + \frac{\sin 3\pi x \cos 6\pi t}{3^3} + \frac{\sin 5\pi x \cos 10\pi t}{5^3} \right.$$

$$\left. + \frac{\sin 7\pi x \cos 14\pi t}{7^3} + \cdots \right) + \frac{1}{2\pi} \sin \pi x \sin 2\pi t$$

The required values are given in Figure 5.2.2

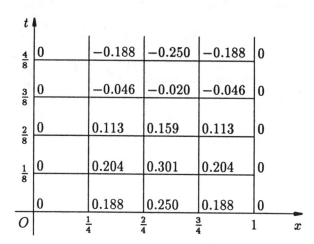

FIGURE 5.2.2

In comparison with Figure 4.4.8, our results correspond with those obtained by the finite-difference method there exactly for $t_0 = 0$ and $t_4 = \frac{4}{8}$ only. Otherwise it is not the case. (The main reason lies not only in application of different methods but also in use of formula (4.2.19) which is applicable for finer time steps only.)

5.2.5—The Fourier method, as presented here, requires zero boundary conditions

$$u(0, t) = 0$$

$$u(l, t) = 0$$

How should we treat problem

$$\frac{\partial u}{\partial t} = a^2 \frac{\partial^2 u}{\partial x^2} \quad \text{in } \Omega \quad (0 < x < l, \quad t > 0)$$

$$u(x, 0) = g(x)$$

$$u(0, t) = b$$

$$u(l, t) = c$$

where $b \neq 0$ and $c \neq 0$ are some constants?

Hint: Use the substitution

$$u(x, t) = v(x, t) + b + \frac{c - b}{l} x \tag{5.2.2}$$

The new unknown function v then satisfies

$$\frac{\partial v}{\partial t} = a^2 \frac{\partial^2 v}{\partial x^2}$$

$$v(x, 0) = g(x) - b - \frac{c - b}{l} x$$

$$v(0, t) = 0$$

$$v(l, t) = 0$$

The original solution u is then obtained by Equation (5.2.2).

5.2.6—By using the same idea, solve the problem

$$\frac{\partial^2 u}{\partial t^2} = a^2 \frac{\partial^2 u}{\partial x^2} \quad \text{in } \Omega \quad (0 < x < l, \quad t > 0)$$

$$u(x, 0) = g(x)$$

$$\frac{\partial u}{\partial t}(x, 0) = h(x)$$

$$u(0, t) = b + ct$$

$$u(l, t) = m + nt$$

Hint: Use the substitution

$$u(x, t) = v(x, t) + b + ct + \frac{(m + nt) - (b + ct)}{l} x$$

then

$$\frac{\partial^2 v}{\partial t^2} = a^2 \frac{\partial^2 v}{\partial x^2}$$

$$v(x, 0) = g(x) - b - \frac{m - b}{l} x$$

$$\frac{\partial v}{\partial t}(x, 0) = h(x) - c - \frac{n - c}{l} x$$

$$v(0, t) = 0$$

$$v(l, t) = 0$$

5.2.7—(Derivatives in boundary conditions) Write the solution of the problem

$$\frac{\partial u}{\partial t} = a^2 \frac{\partial^2 u}{\partial x^2} \qquad (5.2.3)$$

$$u(x, 0) = g(x) \qquad (5.2.4)$$

$$\frac{\partial u}{\partial x}(0, t) = 0 \qquad (5.2.5)$$

$$\frac{\partial u}{\partial x}(l, t) = 0$$

Hint: We have, in $[0, l]$ (see, e.g., K. Rektorys 1994, [1] Section 16.3)

$$g(x) = \frac{a_0}{2} + \sum_{n=1}^{\infty} a_n \cos \frac{n\pi x}{l}$$

where

$$a_n = \frac{2}{l} \int_0^l g(x) \cos \frac{n\pi x}{l} \, dx \qquad (5.2.6)$$

Thus

$$u(x, t) = \frac{a_0}{2} + \sum_{n=1}^{\infty} a_n e^{-\frac{a^2 \pi^2 n^2}{l^2} t} \cos \frac{n\pi x}{l} \qquad (5.2.7)$$

Note: for $t \to \infty$ each term on the right-hand side of Equation (5.2.7) tends to zero except the first one, which is equal to

$$\frac{1}{l} \int_0^l g(x) \, dx \qquad (5.2.8)$$

This result corresponds well to the physical meaning of the problem: Boundary conditions (5.2.5) mean that the heat cannot escape outside, so that the "asymptotic temperature" (5.2.8) corresponds to the "initial heat content," proportional to $\int_0^l g(x) \, dx$.

5.2.8—Consider the problem

$$\Delta^2 u \equiv \frac{\partial^4 u}{\partial x^4} + 2 \frac{\partial^4 u}{\partial x^2 \partial y^2} + \frac{\partial^4 u}{\partial y^4} = \frac{1}{D} g(x, y) \text{ in } \Omega \qquad (5.2.9)$$

$$u = 0 \qquad (5.2.10)$$

$$\frac{\partial^2 u}{\partial v^2} = 0 \quad \text{on } \Gamma$$

where

$$\Omega = (0, a) \times (0, b)$$

is a rectangular region with the boundary Γ, $\partial^2 u / \partial v^2$ the second-order outward-normal derivative (deflection of a horizontal simply supported plate loaded by a vertical load $g(x, y)$). Show that

    1. If

$$g(x, y) = \sum_{m=1}^{\infty} \sum_{n=1}^{\infty} \alpha_{mn} \sin \frac{m\pi x}{a} \sin \frac{n\pi y}{b}$$

(cf. K. Rektorys 1994, [1] Theorem 16.3.5), then the (formal) solution of problems (5.2.9) and (5.2.10) is

$$u = \frac{1}{\pi^4 D} \sum_{m=1}^{\infty} \sum_{n=1}^{\infty} \frac{\alpha_{mn}}{\left(\frac{m^2}{a^2} + \frac{n^2}{b^2}\right)^2} \sin \frac{m\pi x}{a} \sin \frac{n\pi y}{b}$$

2. If $g(x, y) = q = \text{const}$, then

$$u = \frac{16q}{\pi^6 D} \sum_{m,n=1,3,5,\dots} \frac{\sin \frac{m\pi x}{a} \sin \frac{n\pi y}{b}}{mn \left(\frac{m^2}{a^2} + \frac{n^2}{b^2}\right)^2}$$

Hint: For (1) see Problem (2.2.3); for (2) consult the previously mentioned Theorem 16.3.5 in K. Rektorys 1994, [1].

References

[1] Rektorys, K., *Survey of Applicable Mathematics I, II,* 2nd rev. ed., Dordrecht–Boston–London, Kluwer, 1994.

[2] Rektorys, K., *Variational Methods in Mathematics, Science and Engineering,* 2nd ed., Dordrecht–Boston–London, Reidel, 1980.

[3] Rektorys, K., *The Method of Discretization in Time and Partial Differential Equations,* Dordrecht–Boston–London, Reidel, 1982.

[4] Babuška, I. and Szabó, B., *Finite Element Analysis,* New York, John Wiley & Sons, 1991.

[5] Axelsson, O. and Barker, V.A., *Finite Element Solution of Boundary Value Problems. Theory and Computation,* New York, Academic Press, 1984.

[6] Ciarlet, P.G., *Finite Element Method for Elliptic Problems,* Amsterdam, Elsevier, 1988.

[7] Zienkiewicz, O.C., *The Finite Element Method in Engineering Science,* 3rd ed., London, McGraw-Hill, 1977.

[8] Ženíšek, A., *Nonlinear Elliptic and Evolution Problems and Their Finite Element Approximations,* London, Academic Press, 1990.

[9] Schwarz, H.R., *Methode der finiten Elemente,* Stuttgart, Teubner, 1984.

[10] Křížek, M. and Neittaanmäki, P., *Finite Element Approximation of Variational Problems and Applications,* Harlow, Longman, 1990.

[11] Whiteman, J.R., *A Bibliography for Finite Elements,* London, Academic Press, 1975.

[12] Marsal, D., *Finite Differenzen und Elemente,* Berlin, Springer, 1988.

Index